Greenhouse Pest Management

Greenhouse Pest Management

RAYMOND A. CLOYD

CRC Press
Taylor & Francis Group
Boca Raton London New York

CRC Press is an imprint of the
Taylor & Francis Group, an **informa** business

CRC Press
Taylor & Francis Group
6000 Broken Sound Parkway NW, Suite 300
Boca Raton, FL 33487-2742

First issued in paperback 2020

ISBN 13: 978-0-367-57477-2 (pbk)
ISBN 13: 978-1-4822-2778-9 (hbk)

Library of Congress Cataloging-in-Publication Data

Names: Cloyd, Raymond A., author.
Title: Greenhouse pest management / Raymond A. Cloyd.
Other titles: Contemporary topics in entomology series.
Description: Boca Raton, FL : CRC Press, Taylor & Francis Group, 2016. |
Series: Contemporary topics in entomology series | Includes
bibliographical references and index.
Identifiers: LCCN 2015041134 | ISBN 9781482227789 (alk. paper)
Subjects: LCSH: Insect pests--Control. | Mites--Control. | Greenhouse
plants--Diseases and pests--Control.
Classification: LCC SB936 .C56 2016 | DDC 631.5/83--dc23
LC record available at http://lccn.loc.gov/2015041134

Visit the Taylor & Francis Web site at
http://www.taylorandfrancis.com

and the CRC Press Web site at
http://www.crcpress.com

Contents

Preface

Greenhouses provide opportunities to grow many different horticultural crops throughout the year. However, greenhouse producers typically deal with a multitude of insect and mite pests during the production of greenhouse-grown horticultural crops. Therefore, greenhouse producers require information on how to properly identify and manage specific insect and mite pests. Despite the availability of a number of publications associated with pest management in greenhouses, there is a need for a comprehensive publication that contains relevant information, images of the major insect and mite pests for identification, and a complete assessment of the management strategies that can be implemented to deal with insect and mite pests.

This book, *Greenhouse Pest Management*, was developed and designed to address these needs. This comprehensive book contains nine chapters (including an Introduction and Suggested Readings) with specific chapters devoted to topics affiliated with greenhouse pest management. The content of each chapter is designed to provide information that is applicable in commercial greenhouse production systems. This book contains technical information and is amply illustrated with images of the major insect and mite pests and subsequent damage they cause to greenhouse-grown horticultural crops. The book also describes, in detail, management strategies including scouting, cultural control and sanitation, physical control, pesticides, and biological control with pertinent representative images.

Greenhouse Pest Management serves as a practical reference and resource for greenhouse producers, extension agents, crop consultants and advisors, students, and researchers. There is a "Suggested Readings" section that includes both scientific peer-reviewed and non-peer–reviewed publications that can be accessed by individuals wanting more information on specific topics. Comments regarding the usefulness of this book are most welcome. Any suggestions will be helpful in increasing the quality and practicality of future editions.

About the Author

Raymond A. Cloyd is a Professor and Extension Specialist in Horticultural Entomology/Plant Protection, Department of Entomology, Kansas State University.

He has an extension (70%) and research (30%) appointment at Kansas State University (Manhattan, Kansas). His research and extension program involves pest management in greenhouses, nurseries, landscapes, turfgrass, conservatories, interiorscapes, Christmas trees, and vegetables and fruits. He is the extension specialist in horticultural entomology for the state of Kansas with a major clientele that includes homeowners, master gardeners, and professional and commercial operators. He has published over 70 scientific refereed publications and over 500 trade journals on topics related to pest management/plant protection. In addition, he has authored or coauthored numerous books (*Pests and Diseases of Herbaceous Perennials*, *IPM for Gardeners*, *Plant Protection: Managing Greenhouse Insect and Mite Pests*, and *Compendium of Rose Diseases and Pests*), book chapters, manuals, picture or pocket guides, and extension-related publications. He is a frequent speaker at state, national, and international conferences and seminars. He has received numerous awards and honors, including the American Society for Horticultural Sciences Outstanding Extension Educator Award 2012; 2011 Society of American Florists Alex Laurie Award for Research and Extension; 2010 Entomological

Society of America North Central Branch Award of
Excellence in Integrated Pest Management; 2008 Epsilon
Sigma Phi, Alpha Rho State Early Career Extension Award;
2003 College of Agricultural, Consumer, and Environmental
Sciences Faculty Award for Excellence in Extension; and
2003 Visionary Leadership Award from Epsilon Sigma Phi
Extension Fraternity.

Chapter 1

Introduction

Greenhouses may be glass or plastic (Figure 1.1), and are designed for growing one or multiple horticultural crops continuously throughout the year due to the ideal environmental conditions such as temperature, relative humidity, and light intensity that are favorable for plant growth and development (Figures 1.2 through 1.4). However, these same environmental conditions also favor the development and reproduction of insect and mite pests, allowing for multiple generations to occur during the growing season. Sustained development of pest populations impact: (1) the frequency of pesticide applications, (2) the importance of rotating pesticides with different modes of action, and (3) the feasibility of using biological control. Furthermore, there is usually continuous production in greenhouses, which is typically associated with high plant densities (Figure 1.5) that can impact the effectiveness of pest management strategies. However, the low temperatures and light levels that occur in winter through early spring may diminish the activity of many insect and mite pests, although this is contingent on geographic location.

Horticultural crops grown in greenhouses are typically well irrigated and fertilized. These ideal growing conditions result in plants that are an attractive food source for many insect and mite pests (Figure 1.6). Consequently, greenhouse-grown

Figure 1.1   Glass greenhouse structures.

Figure 1.2   Greenhouse production of multiple crops on benches.

horticultural plants can provide an abundance of food for insect and mite pests, often forcing greenhouse producers to implement strategies that reduce plant damage and maintain crop marketability. The types of horticultural crops grown in greenhouses include annuals, perennials, tropicals (foliage),

Figure 1.3   Greenhouse production of multiple crops on floor.

Figure 1.4   Greenhouse production of multiple crops in hanging baskets and liners.

woody plant material, orchids, and herbs and vegetables (Figure 1.7). In addition, many different crops, varieties, and cultivars are grown together. Greenhouse-grown horticultural crops are typically sold to florists, independent garden centers, or chain stores. The different crops, varieties, and cultivars can

Figure 1.5 **Multiple crops being grown in greenhouse.**

Figure 1.6 **Different plant types grown simultaneously in the same greenhouse.**

vary in their susceptibility to certain insect and mite pests, thus impacting the input of pesticides and potential for plant injury (such as phytotoxicity). Nonetheless, multiple crops growing at different stages of development make pest management a challenge.

Figure 1.7   **Greenhouse production of annuals, herbs, and vegetables.**

Insect and mite pests can be introduced into greenhouses by the movement of infested plant material from overseas or offshore facilities, between states, and even within greenhouses associated with the same facility. In addition, movement of infested plant material either within or among greenhouses can contribute to spreading insect and mite pests to noninfested crops. Insect and mite pests can be distributed within a greenhouse by horizontal air-flow fans or workers, or by the movement of already infested plant material from one greenhouse to the next. Greenhouses can also isolate insect or mite pest populations from natural enemies, such as parasitoids and/or predators that could potentially enter from outside.

Greenhouse-grown vegetable crops (Figure 1.8) or cut flowers (Figure 1.9) can tolerate higher levels of insect and mite pest populations, and a certain level of damage to the foliage may be acceptable because only the fruits and flowers are sold. However, nearly all ornamental crops are sold for their aesthetic value, with the flowers and foliage sold as a unit, so the tolerance level associated with insect or mite pest

Figure 1.8   Tomato production in greenhouse.

Figure 1.9   Production of transvaal daisy cut flowers.

damage is low. Consequently, greenhouse producers apply pesticides on a routine basis during the growing season, which places undo selection pressure on pest populations. Moreover, both ornamental and vegetable crops may be present in the same greenhouse, thus impacting pest management decisions. Similarly, pest management is difficult due to the continuous production of multiple crops (Figure 1.10). This results in greenhouse producers typically encountering multiple pest complexes in which many crops are attacked simultaneously by several insect and/or mite pests. As such, pesticides are applied on a frequent basis. In general, pest management comprises approximately ≤5% of the overall costs associated with plant production in most greenhouse operations. Therefore, as long as a given pesticide is effective, despite the cost, greenhouse producers will purchase the product.

Currently, greenhouse producers receive plants from offshore suppliers in Central or South America that frequently apply pesticides (e.g., insecticides and miticides), including some that may not be labeled for use in the United States, in order to maintain plant quality (Figure 1.11). However, insect and mite pests can be preconditioned to develop pesticide resistance on plants that are then used by the recipient greenhouse producer. The movement of plant material within different countries also exacerbates the dispersal of insect and mite

Figure 1.10   **Production of multiple crops in greenhouse.**

Figure 1.11   Application of pesticide in open greenhouse (sides rolled up).

pests, especially those with life stages that include winged adults (e.g., thrips, whiteflies, and leafminers). Therefore, pest management in greenhouses can be quite challenging.

This book was developed to provide information on pest management in greenhouse production systems with designated chapters on insect and mite pest identification, scouting, cultural control and sanitation, physical control, pesticides, and biological control.

Chapter 2

Pest Identification

There are a number of insect and mite pests that may be encountered during the process of growing horticultural crops in greenhouses, although the types will vary depending on geographic location and plant species grown. Insect and mite pests of greenhouse-grown horticultural crops have two primary types of feeding behaviors: piercing–sucking or chewing. Insects with piercing–sucking mouthparts feed within the phloem sieve tubes to obtain free amino acids (e.g., building blocks for proteins) that are essential for development and reproduction. In order to acquire the necessary quantities of amino acids, insects must consume large quantities of plant fluids, which contain an assortment of other materials in higher quantities than amino acids. Subsequently, the excess is excreted as honeydew, which is a clear sticky liquid that accumulates on plant parts such as leaves and stems. Insects with chewing mouthparts consume plant tissue (e.g., leaves, stems, flowers, or roots) during feeding, resulting in immediate and noticeable damage to plants.

Insect and mite pests may be present on aboveground plant parts such as leaves, buds, flowers, and/or fruit or belowground in the growing medium. Insect and mite pests can cause direct plant damage by feeding on plant tissues.

The direct damage symptoms, based on feeding type, associated with most of the insect and mite pests of greenhouse-grown horticultural crops are presented in Table 2.1. However, some insect pests may also cause indirect damage due to the presence of by-products such as molting skins, fecal deposits (frass), and honeydew. Furthermore, certain

Table 2.1 Feeding Types, Pests, and Associated Damage Symptoms for the Different Insect and Mite Pests of Greenhouse-Grown Horticultural Crops

Feeding Type	Pests	Damage Symptoms
Phloem feeders	Aphids, whiteflies, mealybugs, and soft scales	Stunting, wilting, leaf distortion, and leaf yellowing (lower leaves).
Chewers	Caterpillars and fungus gnat larvae	Removal of plant tissues including leaves, stems, flowers (caterpillars), and roots (fungus gnat larvae). Wilting, stunting, and leaf yellowing (fungus gnat larvae).
Miners	Leafminers	Serpentine or blotched mines in leaf tissue.
Chlorophyll feeders	Twospotted spider mite	Leaves appear "stippled," "speckled," or "bronzed," and webbing may be present.
Mesophyll or epidermal fluid feeders	Western flower thrips	Leaves have a "silvery" appearance with sunken tissues on leaf undersides. Black fecal deposits may be present on the undersides of leaves.
Internal feeders	Broad and cyclamen mites	Stunted or distorted growth and small leaves. Damage may resemble nutrient deficiencies or viruses.

insect pests such as thrips, aphids, and whiteflies can vector viruses. Pests that transmit viruses present a formidable challenge in regards to pest management. Many insect and mite pests feed on a multitude of different greenhouse-grown horticultural crops, and they even prefer feeding on certain plant types (e.g., cultivars or varieties) either due to plant chemistry or nutritional content, which can influence their distribution in a greenhouse.

The primary insect and mite pests of greenhouse-grown horticultural crops include aphids, fungus gnats, leafminers, mealybugs, mites (e.g., twospotted spider, broad, and cyclamen), shore flies, thrips, and whiteflies. There are also a number of minor pests, some not insect or mite related, that may be a problem or encountered during the course of the growing season such as caterpillars, leafhoppers, scales, snails and slugs, and sowbugs/pillbugs.

Aphids

Aphids are soft-bodied insects approximately 3.1 mm (1/8 in.) long that can vary in color from yellow, green, black, orange, brown to pink (Figures 2.1 and 2.2). Color varies depending on the plant type fed upon. Therefore, color should not be used to identify aphids. There are a number of different aphid species that attack greenhouse-grown horticultural crops; however, the two most commonly encountered species in greenhouses are the green peach aphid (*Myzus persiae*) and cotton/melon aphid (*Aphis gossypii*). Additional aphid species that can be present in greenhouses, depending on crops being grown, are the potato (*Macrosiphum euphorbiae*) and foxglove (*Acyrthosiphon solani*) aphid.

All aphids possess tubes on the base of the abdomen called cornicles (Figure 2.3), which is where alarm pheromones are emitted when predators, such as ladybird beetles, are present.

Figure 2.1   Aphids on leaf underside.

Figure 2.2   Green peach aphids on underside of leaf.

In greenhouses, all aphids are females that give birth to live offspring, and are capable of reproducing in 7–10 days. Aphids do not have to mate to reproduce, which is referred to as *parthenogenesis*. Aphids reproduce over a 20–30-day period. One female aphid can give birth to approximately 100 live nymphs.

Figure 2.3   Close-up of aphid feeding.

Most adult aphids feeding on plants are wingless; however, winged adults (Figure 2.4) will develop within a population when plant nutritional quality declines or when plants become crowded with too many aphids, thus allowing aphids to disperse and locate a new food source within the greenhouse.

Figure 2.4   Winged aphid adults.

Aphids can develop and reproduce throughout the year in greenhouses.

Aphids have piercing–sucking mouthparts that allow them to feed in the phloem sieve tubes withdrawing plant fluids. Aphid feeding can result in plant stunting, leaf yellowing, and distorted plant growth. In addition to direct feeding damage, aphids can also cause indirect damage by vectoring viruses. During the feeding process, aphids excrete a clear, sticky liquid called honeydew (Figure 2.5) that serves as a substrate for certain black sooty mold fungi (Figure 2.6). Black sooty mold can cover the leaf surface and inhibit the plants ability to manufacture food by means of photosynthesis. Furthermore, ants feed on the honeydew (Figure 2.7), and will protect aphids from natural enemies, including parasitoids and predators. Aphids will also leave behind white cast skins after molting (Figure 2.8). The cast skins may resemble whitefly adults, so be sure to observe closely using a 10× hand lens.

Figure 2.5   **Honeydew on leaf surface.**

Figure 2.6   **Black sooty mold on leaves.**

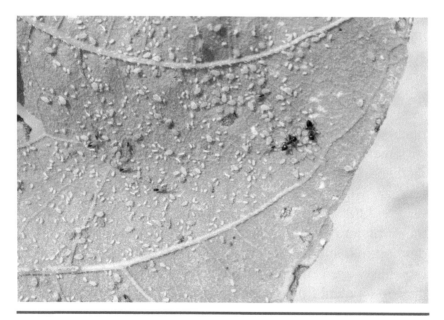

Figure 2.7   **Ants tending aphids on leaf surface.**

Figure 2.8 Aphid molting skins on leaf.

Broad Mite

Broad mite (*Polyphagotarsonemus latus*) adults are approximately 0.25 mm (0.0009 in.) in length, shiny, oval shaped, and amber to dark green in color (Figure 2.9). There are four life stages: egg, larva, nymph, and adult. Females are capable of laying up to 40 eggs during their approximate 2-week life span; however, longevity is contingent on ambient air temperature and relative humidity. Unmated females produce only males, and sons of virgin females can mate with their mothers producing eggs that hatch into female offspring. Eggs are white and oval shaped with bumps or protrusions (Figure 2.10). Six-legged larvae emerge from eggs and then transition into eight-legged nymphs. Nymphs eventually develop into adults. Broad mite females have short, thin hind legs. Males are generally smaller than females.

Development from egg to adult takes 5–6 days at 21°C–26°C (70°F–80°F) and 7–10 days at 10°C–18°C (50°F–65°F). Broad mites are primarily spread within a greenhouse by air currents such as those created by horizontal air-flow fans, leaves

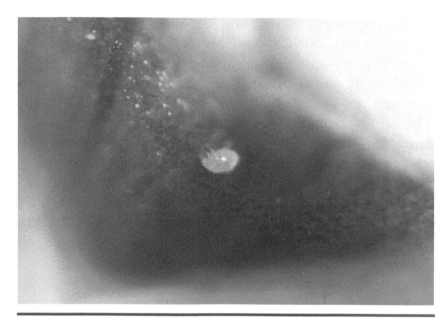

Figure 2.9   Close-up of broad mite adult.

Figure 2.10   Broad mite adult and egg (note bumps or protrusions on surface of egg).

of adjacent plants touching, and workers handling infested plants and then touching noninfested plants. Depletion of a food source associated with low nutritional quality may exacerbate broad mite dispersal among greenhouse-grown horticultural crops because broad mites will search for plants with a higher nutritional content. In addition, broad mite females can attach themselves to the legs and antennae of greenhouse whitefly (*Trialeurodes vaporariorum*) and sweet potato whitefly (*Bemisia tabaci*) adults. However, whitefly adults may not remain stationary long enough for the mites to attach. Broad mites do not appear to attach to thrips or aphids. Male broad mites play an active role in distribution by carrying female nymphs to young leaves, and they can also transport eggs and female adults to new leaves.

Broad mites tend to aggregate in groups when feeding, mainly on the underside of leaves and in flowers where females lay eggs. Broad mites feed on plant cells within the leaf epidermis using their piercing–sucking mouthparts. Their feeding causes leaf bronzing, leaf margins to curl downward and become brittle, and puckered and shriveled growth (Figure 2.11).

Figure 2.11   **Broad mite feeding damage on pepper plant.**

Figure 2.12   Broad mite feeding damage on poinsettia plant.

In addition, broad mites inject toxins during the feeding process. Extensive broad mite populations can migrate and feed on the upper side of leaves causing severe leaf distortion. Broad mite feeding damages the meristematic tissue of the growing tip or apical shoot, thus inhibiting growth, decreasing leaf number, leaf size, and leaf area, and subsequently reducing plant height (Figure 2.12). Furthermore, leaves may appear firm and darker green than normal (Figure 2.13). Broad mites can also cause distortion or malformation of flowers. Feeding damage caused by broad mite may resemble exposure to a phenoxy-based herbicide (e.g., 2,4-D), a virus, or nutritional imbalance such as magnesium or boron deficiency (Figure 2.14).

Caterpillars

Caterpillars are the larval or immature stage of butterflies or moths. Caterpillars can be a problem from late spring through fall depending on geographic location. Adults enter greenhouses through openings such as vents, doors, and

Figure 2.13   Broad mite feeding damage on transvaal daisy plant (note darkened-purplish color of new leaves).

Figure 2.14   Broad mite feeding damage on garden impatiens.

sidewalls. Females lay eggs on plant leaves that hatch into caterpillars with chewing mouthparts. A number of caterpillars may be encountered feeding on greenhouse-grown horticultural crops including the beet armyworm (*Spodoptera exigua*), cabbage looper (*Trichoplusia ni*), corn earworm

(*Helicoverpa zea*), diamondback moth (*Plutella xylostella*), European corn borer (*Ostrinia nubilalis*), imported cabbageworm (*Artogeia rapae*), and tobacco budworm (*Helicoverpa virescens*) (Figure 2.15).

Several caterpillars feed on certain plant types in particular plant families. For example, the cabbage looper, diamondback moth, and imported cabbageworm feed on plants in the cole crop family (Cruciferae). Cabbage looper caterpillars are light green in color, approximately 38.1 mm (1½ in.) in length with white stripes that extend down the back and along the side of the body. In addition, cabbage loopers possess three pairs of legs near the head and three additional pairs (called prolegs) near the last abdominal segment (end of body) (Figure 2.16). Diamondback moth caterpillars are about 12 mm (1/3 in.) long, light green in color, and may either mine or chew plant leaves (Figure 2.17). Imported cabbageworm caterpillars are 31.2 mm (1¼ in.) in length, velvety green in color with a yellow stripe extending down the back and a broken line of yellow spots along each side of the body (abdomen) (Figure 2.18).

Figure 2.15   **Tobacco budworm caterpillar.**

Figure 2.16   **Cabbage looper caterpillar.**

Figure 2.17   **Diamondback moth caterpillar.**

The life cycle includes an egg, larva (caterpillar), pupa, and adult. Adult females are usually active at night although some species may be observed during the daytime. Females lay eggs on the underside of leaves (Figure 2.19). The number of eggs laid will vary depending on the species with females laying

Figure 2.18   **Imported cabbageworm caterpillar.**

Figure 2.19   **Caterpillar eggs on leaf underside.**

20–100 eggs during their lifetime. The eggs hatch into caterpillars that feed on plant leaves. Caterpillars will go through a series of stages referred to as instars and increase in size from one instar to the next. Depending on the species, there may be three to five instars. Initially, caterpillars feed extensively

and then consume less as they prepare to pupate. The caterpillar stage may last 7–10 days although this is contingent on
the species. Caterpillars eventually develop into a pupae (tran-
sitional) stage. Some caterpillars spin cocoons while others
do not. Pupae may be found on plants, on stems, or on the
surface of the growing medium. After 1 week, adults emerge
from the pupae. A typical life cycle, from egg to adult, is
approximately 3–4 weeks although this depends on the plant
fed upon and ambient air temperature.

Caterpillars have chewing mouthparts and cause direct
damage by consuming plant parts including flowers, fruit,
and leaves. They can consume the entire leaf or leave the
midvein (Figure 2.20). Fecal deposits may be present on
plant leaves and stems (Figure 2.21), which is an indica-
tion of caterpillar activity. Some caterpillars will roll leaves
together with silken threads, whereas others will tunnel into
plant stems. If caterpillar populations are left unchecked,
they can cause extensive plant damage that reduces crop
marketability.

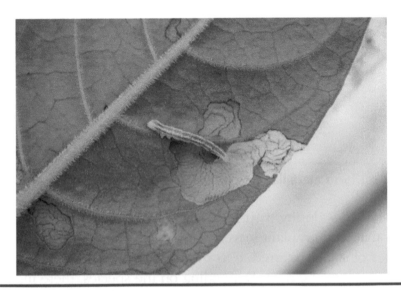

Figure 2.20   **Damage caused by caterpillar feeding.**

Figure 2.21 **Caterpillar fecal deposits on top of leaf.**

Cyclamen Mite

Cyclamen mite (*Phytonemus pallidus*) is very similar to broad mite with respect to biology, life cycle, and feeding damage. Cyclamen mite adults are about 0.25 mm (0.0009 in.) long, oval shaped, yellow to brown in color, and appear transparent. In contrast to broad mite, eggs are oval and smooth with no bumps or protrusions. The life cycle, from egg to adult, takes between 1 and 3 weeks to complete although this is dependent on ambient air temperature. Cyclamen mite females lay between 1 and 3 eggs per day, and up to 16 eggs during their approximate 2-week life span. The eggs are deposited in clusters within terminal buds. Similar to broad mite, cyclamen mite feeds within plant cells causing leaf distortion or twisting, bronzing, and leaf curling. Feeding typically results in leaves appearing wrinkled, brittle, and rough. Plants heavily infested with cyclamen mites will be stunted with small leaves that eventually turn brown to silver. Furthermore, flower buds may abort or not open at all (Figure 2.22).

Figure 2.22 Cyclamen mite feeding damage on cyclamen flower.

Fungus Gnats

Adult fungus gnats (*Bradysia* spp.) are winged, approximately 3–4 mm (0.12–0.16 in.) in length, with long legs and antennae (Figure 2.23). They live from 7 to 10 days. Adults primarily aggregate near the growing medium surface. Females deposit 100–200 eggs into the cracks and crevices of the growing medium that hatch into white, translucent, legless larvae that are about 6 mm (0.23 in.) long. A distinguishing characteristic of fungus gnat larvae is the distinct black head capsule (Figure 2.24).

The life cycle consists of an egg, four larval instars, pupa, and adult, which can be completed in 20–28 days depending on growing medium temperature. Larvae are located within the top 2.5–5 cm (1–2 in.) of the growing medium or inside plant tissue such as the crown or base. Fungus gnat larvae feed on plant roots including the root hairs (Figure 2.25), and organic matter in the upper 2 cm (0.78 in.) of the growing medium. However, larvae may be distributed throughout the growing medium profile, even at the bottom of containers

Figure 2.23   **Fungus gnat adults on yellow sticky card.**

Figure 2.24   **Fungus gnat larvae (note distinct black head capsule).**

Figure 2.25   Root-feeding damage caused by fungus gnat larvae.

near drainage holes. Moreover, fungus gnat larvae may emerge
from the growing medium and feed on leaves lying on the
surface of the growing medium (Figure 2.26) or tunnel directly
into plant crowns. Larvae prefer a moist growing medium, and
require fungi as a supplemental food source in order to complete development.

Abundance and fitness of adults, as well as reproductive capacity of females, are determined by food type. Larval
feeding directly damages developing root systems and inter-
feres with the ability of plant roots to absorb water and nutrients from the growing medium resulting in stunted growth.
Furthermore, larvae may cause indirect damage during feeding by creating wounds that allow entry of soil-borne plant
pathogens such as *Pythium* spp. Fungus gnats may be present
in bagged growing medium, which means that pasteurization
or treating the growing medium with heat may be required to
avoid issues with fungus gnats.

Larvae and adults can also transmit fungal diseases such
as *Pythium* spp., *Fusarium* spp., *Thielaviopsis basicola*, and
Botrytis spp. from infected to noninfected plants. Fungus

Figure 2.26 **Transvaal daisy leaf lying on growing medium surface fed upon by fungus gnat larvae.**

gnat adults can carry the spores of certain foliar and soil-borne plant pathogens on their bodies including *Botrytis cinerea, Fusarium oxysporum* f. sp. *radicus-lycopersici,* and *Thielaviopsis basicola.* Adults may then disperse spores throughout a greenhouse. Fungus gnat larvae can ingest the propagules of *Pythium aphanidermatum* and macroconida of *Fusarium avenaceum,* which are then introduced into young healthy plants during feeding. In addition, the oospores of *Pythium* spp. can survive passage through the digestive tract of fungus gnat larvae and may be intact and viable after being excreted.

Leafhoppers

Adult leafhoppers (in various genera) vary in size from 3.1 to 4.7 mm (1/8–3/16 in.) in length (Figure 2.27). Leafhoppers are slender and wedge shaped with a tapered end (Figure 2.28). They are typically yellow to pale green

Figure 2.27   Leafhopper adult on leaf underside.

Figure 2.28   Leafhopper adult on top of leaf.

in color. Nymphs and adults move sideways on leaves when disturbed. Adults can fly because they have wings, whereas nymphs are wingless. Females lay eggs into leaves or stems that hatch into nymphs. Nymphs resemble adults but are smaller and lack wings (Figure 2.29).

Figure 2.29   **Leafhopper nymphs on leaf underside.**

There are five instars (nymphs) that take about 2 weeks to develop into adults. Adults may live for 1 month or longer. The life cycle, from egg to adult, takes 20–28 days to complete.

Leafhoppers have piercing–sucking mouthparts that are used to withdraw plant fluids. Depending on the species, leafhoppers feed in the xylem (water-conducting tissues), phloem (food-conducting tissues), or both. Leafhoppers destroy plant cells during feeding and can also inject toxins into plants, which disrupt the movement of fluids throughout plant parts. Feeding by leafhoppers can cause *stippling* of plant leaves (Figure 2.30) that resembles damage from the twospotted spider mite (*Tetranychus urticae*). Damage symptoms include leaf distortion, leaf chlorosis, plant stunting, leaf curling, leaf yellowing, and/or leaf browning or necrosis. White, molted skins may be present on the underside of leaves (Figure 2.31).

Figure 2.30 **Leaf damage caused by leafhopper (this damage is referred to as *stippling*).**

Figure 2.31 **Leafhopper molting skins on leaf underside.**

Leafminers

A number of leafminer species may be encountered in green-houses depending on the crops being grown (e.g., ornamentals vs. vegetables) including the vegetable leafminer (*Liromyza sativae*), serpentine leafminer (*Liromyza trifolii*), and pea leafminer (*Liromyza huidobrensis*). Adult females, in general, are 6.3 mm (1/4 in.) long, and black in color with a yellow head (Figure 2.32). Most leafminers have distinct yellow markings on the body. Females use their ovipositor (egg-laying device at the end of the abdomen) to insert eggs into leaf tissue, subsequently creating punctures in leaves that develop into small white spots or "stipples" on the leaf surface (Figure 2.33). Females feed on fluids that exude from the punctures made in leaves. A single female is capable of laying up to 200 eggs during a 2–3-week life span. Eggs hatch in 5–6 days into larvae that are yellow to brown in color. The larvae are 3.1 mm (1/8 in.) in length when fully mature (Figure 2.34).

Larvae tunnel or create serpentine or blotched mines below the epidermal layer of plant leaves (Figures 2.35 and 2.36)

Figure 2.32 **Leafminer adults.**

Figure 2.33   Leafminer female oviposition punctures on leaf surface.

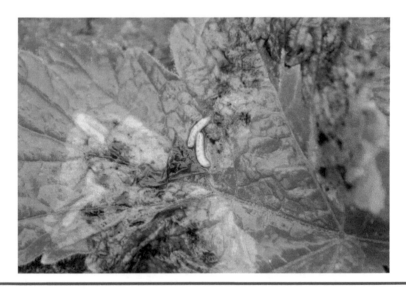

Figure 2.34   Leafminer larvae.

that can negatively affect the aesthetic quality of greenhouse-grown horticultural crops, thus resulting in economic losses. Damage from leafminer larval feeding can reduce the amount of plant leaf area that can photosynthesize (manufacture food). After feeding within the plant leaf tissues for 2 weeks, each

Figure 2.35 Serpentine mines on transvaal daisy leaf caused by leafminer larvae.

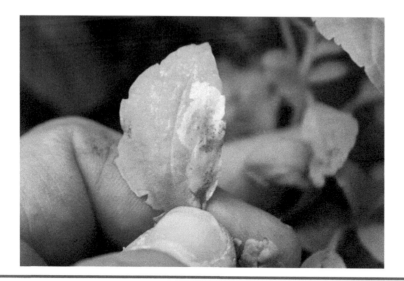

Figure 2.36 Blotched mine on leaf caused by leafminer larva.

fully mature larva chews a hole in the leaf, and drops to the ground to pupate. Adults emerge from the pupae after about 2 weeks and disperse to a new leaf. After mating, females deposit eggs into leaf tissues. Development from egg to adult takes about 5 weeks although development is contingent on the ambient air temperature. Multiple overlapping generations can occur during a cropping cycle.

Mealybugs

Although a number of mealybug species may be encountered during the production of greenhouse-grown horticultural crops, the predominant species is the citrus mealybug, *Planococcus citri*. Citrus mealybugs are oval shaped, segmented with white, waxy protrusions extending around the periphery of the body (Figures 2.37 and 2.38). Females are white, wingless, and 2–5 mm (0.078–0.20 in.) in length when fully grown, whereas males are somewhat smaller. The life cycle consists of five development stages: egg, three nymphs (crawlers), and adult. Males undergo six development stages, which includes two pupal stages: prepupa and pupa. Before female adults die, they lay eggs underneath their body cavity. Eggs hatch into nymphs that actively move around seeking

Figure 2.37 **Citrus mealybugs on plant stem.**

locations to feed. The nymphs are initially yellow-orange in color but eventually turn white after undergoing successive molts (Figure 2.39).

Once nymphs or crawlers find a place to feed, citrus mealybugs progress through several growth stages before becoming

Figure 2.38   Citrus mealybug located at juncture of leaves.

Figure 2.39   Citrus mealybug nymph and molting skin.

adults. Eventually, males become winged individuals that mate with females and then die after 2–3 days. Females continue development and then die after laying eggs. Eggs remain protected underneath the body cavity of the dead female until they hatch. A single citrus mealybug female can lay up to 600 eggs. Only adult males and newly emerged nymphs are able to disperse. Mealybugs, in general, have an extended development period (egg to adult) compared to most other greenhouse insect and mite pests, such as aphids, thrips, whiteflies, and mites. The life cycle from egg to adult takes approximately 60 days; however, development depends on the ambient air temperature and host plant fed upon. The primary ways that mealybug nymphs disperse within a greenhouse are as follows:

1. Wind or air currents
2. Workers handling infested plants and inadvertently transferring mealybugs to uninfested plants
3. Plant leaves touching thus allowing nymphs to migrate among plants
4. Introduction of infested plant material
5. Ants transporting nymphs or crawlers among plants

The lateral waxy protrusions protect mealybugs from natural enemies, such as parasitoids and predators. Mealybugs tend to aggregate in large numbers at leaf junctures where the petiole meets the stem, on leaf undersides, on stem tips, and under the leaf sheaths of certain plants (Figures 2.40 and 2.41). Mealybugs are difficult to detect initially and then suddenly populations become noticeable—resulting in severe outbreaks. When severe outbreaks occur, an effective management strategy will be too late to implement, other than disposing of all infested plants.

Mealybugs cause direct plant damage when feeding on plant fluids in the vascular tissues, primarily the phloem, mesophyll, or both with their piercing–sucking mouthparts.

Figure 2.40   **Mealybugs aggregating on plant stem.**

Figure 2.41   **Citrus mealybugs on stem of poinsettia.**

In addition, mealybugs can inject toxins directly into plants. Feeding may cause leaf yellowing, plant stunting, and wilting. Furthermore, mealybugs excrete a clear sticky liquid called honeydew during feeding, which serves as a substrate for black sooty mold fungi.

Pillbugs and Sowbugs

Pillbugs and sowbugs are not insects or mites but are classified as isopods or crustaceans. Both are oblong, oval or convex in shape, segmented, and are flattened underneath the body. Pillbugs and sowbugs are black, gray, or brown in color, and about 12.7 mm (1/2 in.) long when fully mature, with seven pairs of legs. They are distinctly segmented with seven hardened individual overlapping plates (Figure 2.42). Pillbugs are capable of rolling up into a ball when disturbed (thus the common name, "roly poly"), whereas sowbugs cannot. Sowbugs have two small, tail-like appendages located at the end of the body—pillbugs do not have these appendages. Adults may live up to 2 years or more. Both feed primarily on decaying organic matter and fungi as they have chewing mouthparts; however, if populations become abundant, pillbugs and sowbugs may feed on the stem and/or roots of young seedlings.

Figure 2.42 Pillbug on leaf.

Scales

Scales and mealybugs have similar life cycles and feeding habits. There are two types of scales: soft and hard. Mature female scales vary in size from 3.1 to 6.3 mm (1/8 to 1/4 in.) in length (Figures 2.43 and 2.44). They are wingless and usually legless.

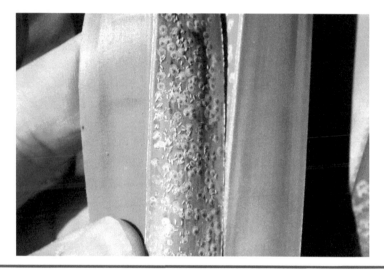

Figure 2.43   **Hard scale on leaf surface.**

Figure 2.44   **Hemispherical scale on leaves.**

Only two life stages are able to move: the first instar (crawler) and the adult male. Eggs laid by females hatch into first instars that have legs, which allow them to move on plant surfaces, such as leaves and stems, searching for a place to feed. The later instars and adult females eventually lose their legs and are unable to move. Females are saclike, wingless, and legless, whereas males develop into winged individuals with legs and one pair of wings. Males do not feed as they lack mouthparts. The females of many scale species are capable of producing live offspring or young while other species may lay eggs.

Eggs are laid underneath the female body and may be hidden or produced in cottony sacs that protrude from the back of the female body. Female scale can lay eggs continually over a 2–3-month period. Soft scale females, in general, lay over 1,000 eggs, whereas hard scale females lay less than 100 eggs. Soft and hard scales vary in color and appearance. Soft scales are oval or globular in shape, whereas hard scales are circular or elliptical in shape. Both soft and hard scales have piercing–sucking mouthparts, which are used to withdraw plant fluids from either the phloem or other plant tissues. Feeding by scales can cause plant wilting, leaf yellowing, and plant stunting. In addition, soft scales like aphids, whiteflies, and mealybugs produce honeydew because the food canal carries large quantities of plant fluids from the phloem sieve tubes. In contrast, hard scales do not excrete honeydew as the food canal of hard scales contains various kinds of cells, and therefore does not transport large quantities of plant fluids. In addition, hard scales do not produce honeydew because they only ingest small amounts of plant fluids. Moreover, they use their long stylets to explore vast areas of plant tissue in order to obtain nutrients for development and reproduction. Ants will move soft scales around, but not hard scales, to prevent soft scales from becoming contaminated by fungi that use honeydew as a substrate. Ants will feed on the honeydew (Figure 2.45), and also protect soft scales from natural enemies including parasitoids and predators.

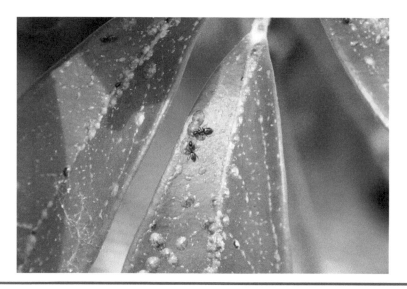

Figure 2.45   Ants tending brown soft scales on schefflera plant.

Shore Flies

Shore fly (*Scatella* spp.) adults resemble houseflies and are 3.1 mm (1/8 in.) long, with black bodies. Each forewing has at least five light-colored spots (Figure 2.46). Antennae and legs are short, and the head is small (Figure 2.47). Larvae are opaque, yellow-to-brown in color, with no black head capsule, and 6.3 mm (1/4 in.) in length. Shore fly adults are stronger fliers than fungus gnat adults. The life cycle, which consists of an egg, three larval stages, pupa, and adult, can be completed in 15–20 days; however, development is contingent on the growing medium temperature.

Shore flies cause minimal direct plant damage compared to fungus gnats. However, shore flies can be a concern because they are more noticeable flying around plants, and are easily detected when captured on yellow sticky cards. Abundant numbers of shore flies can be easily transported from one greenhouse to another through infested plant material. Inadvertently shipping crops with shore fly adults may result in the rejection of plant material shipments. Although

Figure 2.46   Shore fly adults (note white spots on wings).

Figure 2.47   Shore fly adult on leaf of basil plant.

shore fly adults are generally considered a nuisance pest, adults may leave black fecal deposits on plant leaves that can impact a plant's aesthetic quality (Figure 2.48). Shore fly larvae feed primarily on algae and decaying organic matter located on the growing medium surface (Figure 2.49). Larvae

Figure 2.48   Shore fly adult fecal deposits on plant leaves.

Figure 2.49   Algae accumulating on surface of growing medium.

can also be found within the growing medium. However, larvae do not feed directly on plant roots. Similar to fungus gnats, shore fly larvae and adults may transmit soil-borne plant pathogens such as *Thielaviopsis basicola* and *Pythium aphanidermatum*.

Slugs and Snails

Slugs and snails are not insects or mites but are classified as mollusks, which includes clams and oysters. Snails have a hard-shell covering (Figure 2.50) that protects them from extreme environmental conditions (e.g., temperature and sunlight) and natural enemies, whereas slugs do not have a hard outer covering (Figure 2.51). Slugs and snails vary in size from 19 to 50.8 mm (3/4–2 in.) long and may be pale yellow to purple in color, depending on the species. Slugs and snails will lay eggs in the cracks and crevices of moist growing medium or underneath plant containers. Eggs hatch in 10 days or less depending on temperature. Slugs and snails are fully mature within 3 months to 1 year. They both have a radula, which is a tongue-like structure used to scrape plant tissue, thus allowing them to consume seedlings and plant leaves. They create ragged-shaped holes in leaves and stems (Figures 2.52 through 2.54). Slugs and snails are active at night, hiding during the daytime underneath containers, benches, or debris, and they exude a silvery muscus-like fluid as they move around on the surface of the growing medium.

Figure 2.50   Snail on leaf.

Figure 2.51   Slug on growing medium surface.

Figure 2.52   Slug feeding damage on transvaal daisy leaves.

Figure 2.53   **Slug feeding damage on zinnia leaf.**

Figure 2.54   **Slug feeding damage on hosta leaf.**

Twospotted Spider Mite

Twospotted spider mite (*Tetranychus urticae*) adults are 0.3–0.4 mm (0.015–0.020 in.) in length, oval shaped, and can vary in color from yellow-green to red-orange. Adults have distinct black markings on both sides of the body (Figure 2.55). Males are smaller than females, and are more elongated with a pointed abdomen. Female adults live about 30 days and can lay between 100 and 200 small, spherical eggs during a 2-week period. Eggs are deposited on the leaf underside along the midveins. Eggs hatch into yellow-green, six-legged larvae that mature into eight-legged nymphs, and then develop into adults.

Twospotted spider mites are orange-red in color from late summer through fall, enter a diapause (resting) stage in response to shorter daylengths and lower temperatures, and overwinter as fertilized females. Development, from egg to adult, can be completed in 1–3 weeks; however, this depends on the plant fed upon and ambient air temperature. For example, the life cycle takes 14 days at 21°C (70°F)

Figure 2.55   **Close-up of twospotted spider mite adult.**

and 7 days at 29°C (84°F). All life stages are located on leaf undersides because twospotted spider mites are very sensitive to ultraviolet light (sunlight). Twospotted spider mites have piercing–sucking mouthparts, which allow them to feed on individual plant cells, causing damage to the spongy mesophyll, palisade parenchyma, and chloroplasts. Feeding by twospotted spider mites reduces the chlorophyll content in leaves and lowers the plants ability to manufacture food through the process of *photosynthesis*. Damaged leaves are bleached and stippled with small, silver-gray to yellow speckles (Figure 2.56). There can also be fine mottling on the upper leaf surface (Figure 2.57). Heavily infested leaves appear bronzed (Figure 2.58), turn brown, and eventually fall off of plants. Excessive populations and feeding by twospotted spider mites can cause premature defoliation and webbing is usually present (Figure 2.59). The damage symptoms associated with twospotted spider mite feeding can vary depending on the plant type fed upon.

Figure 2.56 **Twospotted spider mite feeding damage on clematis leaf (this damage is referred to as either "specking" or "stippling").**

Figure 2.57   Twospotted spider mite feeding on violet leaves.

Figure 2.58   Twospotted spider mite feeding damage on New Guinea impatiens (note bleaching of leaves).

Figure 2.59 **Twospotted spider mites and webbing on plant (the webbing allows the mites to move onto plant leaves and other plants nearby).**

Western Flower Thrips

Western flower thrips (*Frankliniella occidentalis*) adults are approximately 2 mm (0.078 in.) long, slender, with two pairs of fringed or hairy wings (Figure 2.60). The larvae and adults vary in color from brown to yellow. The life cycle consists of an egg, two larval stages, two pupal stages, and an adult. The life cycle takes 2–3 weeks to complete depending on ambient air temperature with the optimum range between 26°C and 29°C (79°F and 84°F). At optimal temperatures, the life cycle can be completed in 7–13 days. A new generation can occur within 20–30 days, although development depends on ambient air temperature. Females live for approximately 45 days and can lay up to 300 eggs during their lifetime. The eggs are inserted into plant tissues such as leaves and fruits.

Western flower thrips can cause direct damage by feeding on plant leaves, flowers (Figures 2.61 and 2.62), and

Figure 2.60   **Western flower thrips on chrysanthemum flower.**

Figure 2.61   **Western flower thrips larvae on chrysanthemum leaf (this damage is referred to as "silvering").**

fruits. Western flower thrips have piercing–sucking mouthparts; however, they do not directly feed in the phloem sieve tubes like aphids and whiteflies. Instead, they feed within the mesophyll and epidermal cells using a single stylet in the mouth to puncture cells. After inserting their stylet, western

Figure 2.62   Western flower thrips feeding damage on zinnia flower.

flower thrips inject a pair of stylets that lacerate and damage cell tissues. Western flower thrips then ingest fluids from the cells. Symptoms of feeding include leaf scarring, distorted growth (Figures 2.63 through 2.65), sunken tissues on leaf undersides, and/or deformed flowers (Figure 2.66). Leaves and flowers develop a "silvery" appearance because air fills the empty cells after plant fluids have been removed (Figure 2.67). Black fecal deposits may also be present on the underside of leaves (Figure 2.68). Damage to plant leaves and fruits may also be associated with females using their sharpened ovipositor to insert eggs into plant tissue. Furthermore, the wounds created through feeding or oviposition may serve as entry sites for plant pathogenic organisms such as fungi or bacteria.

Western flower thrips cause indirect damage to plants by vectoring the tospoviruses: *impatiens necrotic spot* and/or *tomato spotted wilt* virus. The first and second instar larvae acquire the viruses by feeding on infected weeds or plants, and then adults transmit the viruses to susceptible plants. A plant infected with a virus must be disposed of immediately.

Figure 2.63 Western flower thrips damage on marigold leaves (note distorted leaf growth).

Figure 2.64 Young transvaal daisy leaves damaged by western flower thrips.

Figure 2.65   Western flower thrips feeding damage on pepper plants.

Figure 2.66   Western flower thrips feeding damage on transvaal daisy flower.

The direct and indirect damage caused by western flower thrips can result in an economic loss to greenhouse producers. Western flower thrips populations are difficult to suppress in greenhouse production systems with insecticides for a number of reasons including: (1) small size, (2) broad host range,

Figure 2.67 Chrysanthemum flower damage caused by western flower thrips feeding.

Figure 2.68 Western flower thrips larvae on leaf underside (note sunken tissue and black fecal deposits).

(3) high female reproductive capacity, (4) rapid life cycle, (5) cryptic or thigmotactic behavior, and (6) ability to develop resistance to insecticides. The thigmotactic behavior, in which western flower thrips reside in unopened terminal growth or flower buds, protects larvae and adults from exposure to contact insecticides.

Whiteflies

Whitefly species that can be encountered in greenhouses are the greenhouse whitefly (*Trialeurodes vaporariorum*) and sweet potato whitefly (*Bemisia tabaci*) although this will depend on the horticultural crops grown. Greenhouse whitefly adults are about 4.2 mm (1/6 in.) and sweet potato whitefly adults are 3.2 mm (1/8 in.) long. Both whitefly species have four wings covered with white, waxy powder (Figures 2.69 and 2.70). Greenhouse whitefly adults hold their wings flat over the body, parallel with the leaf surface. Sweet potato whitefly adults retain their wings at a 45° angle, which appear

Figure 2.69 **Whitefly adults on leaf underside.**

Figure 2.70 **Whitefly adults on underside of transvaal daisy leaf.**

rooflike over the body. Female adults can lay up to 20 eggs per day on the underside of plant leaves, and can lay up to 300 eggs during their 30–45 day life span. Eggs of the greenhouse whitefly are pale green to purple in color and eggs of the sweet potato whitefly are white to gray in color with a darkened tip. Eggs hatch within 5–10 days and the newly emerged nymphs (crawlers) search for feeding sites on leaf undersides.

Whitefly nymphs have piercing–sucking mouthparts that are used to withdraw plant fluids from the phloem sieve tubes. Nymphs are typically flattened and transparent to yellow-brown in color. The nymphs do not move after establishing a feeding site and remain immobile for 2–3 weeks, and then they transition into a pupal stage or fourth instar. Greenhouse whitefly pupae have elongated waxy filaments that encircle the outer periphery and are elevated in profile with vertical (perpendicular) sides. The pupae appear cakelike on the leaf surface. Sweet potato whitefly pupae are yellow-brown in color and appear flatlike on the leaf surface with no elongated waxy filaments. Adults emerge from pupae after approximately 1 week. The life cycle, from egg to adult, can be completed in 3–4 weeks depending on the plant type fed upon and ambient air temperature. All life stages are usually located on leaf undersides (Figures 2.71 through 2.73). Adults and nymphs feed on plant fluids causing wilting, leaf yellowing, leaf distortion, and/or plant stunting. Similar to aphids, whiteflies excrete honeydew that accumulates on plant leaves.

Figure 2.71 Adult whiteflies and nymphs on underside of poinsettia leaf.

Figure 2.72   **Heavy infestation of whiteflies on underside of transvaal daisy leaves.**

Figure 2.73   **Whitefly adults and nymphs on underside of leaf.**

Chapter 3

Scouting

Scouting or monitoring is an important component of any pest management program in greenhouse production systems. Furthermore, scouting greenhouse-grown horticultural crops helps prevent outbreaks of insect and mite pest populations from occurring. The goals of a scouting program may consist of one or a combination of the following factors:

1. *Reduce Pesticide Use*: Decreasing the use of pesticides (insecticides and/or miticides) may reduce selection pressure placed on insect and mite pest populations, and lead to a reduction in populations developing resistance. A reduction in pesticide use will allow susceptible individuals to escape mortality and breed with resistant individuals, thus maintaining susceptible genes throughout an insect or mite pest population. Furthermore, a reduction in pesticide use can lead to fewer problems associated with plant injury (such as phytotoxicity) or negatively affecting plant physiology, resulting in healthier plants.

2. *Determine Effectiveness of Management Strategies*: The effectiveness of cultural control and sanitation, physical control, pesticides, and/or biological control strategies must be evaluated to ascertain if current strategies are

maintaining pest populations below damaging levels. The tracking of insect and/or mite pest populations throughout the growing season will determine if any adjustments are necessary to existing pest management programs. In addition, proper record keeping will greatly increase the prospects of timely implementation of the appropriate pest management strategy.

3. *Evaluate Population Dynamics and Trends throughout the Growing Season*: Assessing the population dynamics or numbers of insect and mite pests will determine seasonal abundance and fluctuations in pest populations during the growing season. Again, proper record keeping is imperative in assessing if, and/or, when pesticide applications are necessary.

4. *Improve Pesticide Use*: Timing of pesticide applications is essential to maximize effectiveness against insect and mite pests. Greater mortality can be achieved by making applications when a high percentage of a given insect or mite pest population is in susceptible life stages, such as the immature (nymphs or larvae) or adult.

A number of supplies or materials are required to initiate a proper scouting program based on the four goals described previously. Scouting and record-keeping materials include colored sticky cards (yellow or blue), potato wedges (for fungus gnat larvae), a 10×–16× hand lens, a clipboard, colored flags, data sheets, and a map of each greenhouse. Data obtained from the scouting process may be computerized and incorporated into a spreadsheet such as Microsoft Excel or a similar database (Figure 3.1). Once the required materials for the scouting program are acquired, it is important to understand the scouting techniques used to determine the number of insect and mite pests in a greenhouse. Scouting techniques may be categorized as either *passive* or *active*.

Passive scouting techniques involve the use of traps, such as colored sticky cards (yellow or blue), and potato wedges or

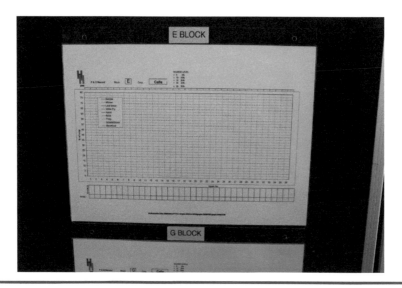

Figure 3.1 Data sheet with information on number of insect pests present per week.

sticks that attract or lure certain insect pests. Colored sticky cards (Figures 3.2 and 3.3) are used to attract and capture the adult stages of fungus gnats, leafminers, shore flies, western flower thrips, whiteflies, and winged aphids. Yellow or blue sticky cards (76.2 mm × 127 mm [3 in. × 5 in.]) may be used. Yellow sticky cards attract a wide variety of flying insect pests, including western flower thrips, which are generally easier to detect or differentiate from other objects on yellow sticky cards, including growing medium (Figure 3.4). Blue sticky cards are generally used when western flower thrips is the dominant insect pest attacking greenhouse-grown crops (Figure 3.5).

For most insect pests, sticky cards should be positioned just above the crop canopy and set by attaching the sticky card to a bamboo stake with a clothespin (Figure 3.6), so that the sticky card can be adjusted as the crop increases in size (height). However, for fungus gnat adults, yellow sticky cards should be positioned horizontally near the growing medium surface because this is where adults are most active. Yellow sticky cards can also be placed on the rims of flats or

Figure 3.2   Yellow sticky card placed among garden impatiens crop.

Figure 3.3   Blue sticky card placed among crop to capture western flower thrips adults.

containers (Figure 3.7). One side of a yellow sticky card can be used for the first week, leaving the protective wax paper on the unused side of the sticky card for the second week. This allows for one yellow sticky card to be used for 2 weeks. When pest numbers are low due to seasonality or when fewer

Figure 3.4 **Yellow sticky card placed above canopy of pansy crop.**

Figure 3.5 **Blue sticky card positioned among crop.**

crops are being grown, sticky cards can be changed less often. Scout at least once per week and record the number of insects detected on the sticky cards on data sheets. The number of sticky cards to place within a crop varies with the greenhouse operation; however, the general rule is one to two sticky cards

Figure 3.6   Yellow sticky card with fungus gnat adults.

Figure 3.7   Yellow sticky card positioned near the growing medium surface to capture fungus gnat adults.

per 46.4–92.9 m^2 (500–1,000 ft.^{2}), although more sticky cards may be needed if the crop is susceptible to the viruses transmitted by the western flower thrips.

In addition to placing sticky cards within a crop, sticky cards can be placed near greenhouse openings such as doors, vents, and sidewalls, which will help detect the migration of winged insects from outside the greenhouse. Sticky cards should also be positioned among the crop near openings such as doors and vents. The placement of sticky cards near greenhouse openings will help detect the adult stages of certain insect pests, including aphids, leafminers, western flower thrips, and whiteflies dispersing inside from weeds, field, and/ or vegetable crops growing nearby. Weeds located outside the greenhouse (Figure 3.8) are a source of insects migrating into the greenhouse and should be removed. Sticky cards should also be placed underneath benches in greenhouses with gravel or soil-based floors (Figure 3.9) to detect the presence of adult fungus gnats, shore flies, and western flower thrips, and help determine if these insect pests are pupating in the soil or gravel underneath benches.

Figure 3.8   **Weeds growing outside of greenhouse.**

Figure 3.9   Yellow sticky card positioned underneath bench to scout for western flower thrips, fungus gnats, or shore flies that may have pupated in soil.

Potato wedges or sticks can be used to detect the presence of fungus gnat larvae (Figure 3.10). For potato wedges, cut potatoes into 6.3 mm (1/4 in.) pieces that are firmly inserted into the growing medium and allowed to remain for 48 hours (Figure 3.11). Reports indicate that allowing the potato wedges to remain on the growing medium surface for 48 hours is more efficient in recovering fungus gnat larvae than 24 hours. Each potato wedge is then removed and the side that was in contact with the growing medium is inspected for the presence of fungus gnat larvae feeding on the potato wedge. Count the number of larvae on the wedges and record the number on a data sheet. For potato sticks, cut potatoes into pieces 7.6–12.7 cm (3–5 in.) in length, and 6.4 mm (1/4 in.) wide. The sticks are inserted into the growing medium, leaving approximately 6.3 mm (1/4 in.) visible. Potato sticks are effective in scouting for fungus gnat larvae located deep within the growing medium and when plants have extensive root systems that reach the bottom of the container. Furthermore, potato sticks may be useful in detecting fungus

Figure 3.10 Potato wedge and sticks used to scout for fungus gnat larvae.

Figure 3.11 Potato wedge placed on growing medium surface to detect the presence of fungus gnat larvae.

gnat larvae in bulb crops, such as Easter lily (*Lilium longiflorum*). After 48 hours, potato sticks are removed, the number of fungus gnat larvae counted, and the number recorded on a data sheet.

Active scouting techniques include visual inspection, that is, looking for insect and mite pests on the underside of leaves or inspecting entire plants (Figure 3.12). For example, a predetermined number of plants (e.g., 20 plants), in a specified area within the greenhouse, such as on a bench, may be randomly selected within a greenhouse section or crop growing area. These plants serve as indicators, are marked or flagged, and used to determine the level of pest infestation (based on numbers of pests present). The number of pests associated with each indicator plant is recorded weekly taking into account plant growth, flowering, and environmental conditions such as temperature, which can influence the number of pests at any given time. For crops that are highly susceptible to the viruses vectored by the western flower thrips, more intensive scouting efforts are required such as checking plants and/or sticky cards twice per week. In order to save time and labor costs,

Figure 3.12   **Visual inspection of poinsettia crop.**

scouting efforts may be concentrated on crops that are highly susceptible to certain insect and/or mite pests. In addition to scouting different susceptible crop types, various cultivars should be checked because particular cultivars of horticultural crops are more susceptible to insect and/or mite pests than others. Areas near openings, including doors and vents, should be scouted more frequently as these are locations where winged insects such as adult leafminers, western flower thrips, and whiteflies are most likely to enter the greenhouse from outdoors. Furthermore, active scouting techniques are effective in detecting nonflying pests such as western flower thrips larvae and whitefly nymphs, young and wingless aphids, female mealybugs, scales, and twospotted spider mites. Another method of scouting (often referred to as the "beat method") involves shaking plant leaves or flowers over a white sheet of paper (21.6 cm × 27.9 cm [8.5 in. × 11 in.]) and counting the number of fallen individuals moving around (Figure 3.13). The "beat method" works well for detecting populations of the twospotted spider mite and western flower thrips.

Figure 3.13 **"Beat method" used to scout for twospotted spider mites and western flower thrips.**

Action Thresholds

Action thresholds are the level or number of insect and/or mite pests that warrant the need to implement measures to avoid problems. The proposed idea associated with an action threshold is that pests can be tolerated at some level, which will reduce the frequency of pesticide applications and consequently pesticide resistance. However, most information is directly related to agricultural cropping systems. Action thresholds may be useful in: (1) determining if management practices are required, (2) timing pesticide applications, and (3) monitoring pest population trends throughout the growing season. The use of action thresholds may lead to a reduction in pesticide inputs. Action thresholds are based on insect and mite counts, or visible plant injury. However, action thresholds will only be helpful if a proper scouting program is established and implemented, which primarily involves using colored (yellow or blue) sticky cards, although under some circumstances visual inspections or using the "beat method" (described previously) may be more appropriate. For example, the "beat method" can provide an accurate assessment of the changes in the population dynamics of both adult and larval stages of the western flower thrips in greenhouse-grown vegetables.

Action thresholds may vary depending on the growing season. For example, higher sticky card counts will most often occur during spring through early fall when insect and mite pests are more active. Furthermore, management practices are intensified, which means that substantially more sticky cards may be required in order to determine an initial infestation. The use of more sticky cards may be effective in suppressing pest populations, thus leading to less damage to horticultural crops.

Action thresholds for insect and mite pests of greenhouse-grown horticultural crops are limited primarily because entire plants are sold, so minimal plant damage can be tolerated.

Moreover, greenhouse producers need to evaluate individual sticky card counts instead of using the average number per sticky card for an entire greenhouse or section. Information associated with the average number per sticky card may be useful in some respects. However, the information may not identify areas with abundant insect and/or mite pest numbers, which would allow greenhouse producers to conduct localized pesticide applications as opposed to making costly broadscale pesticide applications. Furthermore, separate sticky card counts might allow greenhouse producers to assess where the highest level of pest activity is concentrated and where pests are entering the greenhouse.

There are a variety of factors that can confound or result in misleading counts on sticky cards:

1. Some plants may be more attractive to insect pests due to variability in cultivar susceptibility, which may lead to higher numbers of adult insect pests on sticky cards associated with attractive plants.
2. The presence of flowers may result in misleading sticky card counts, especially when scouting for adult western flower thrips because they tend to inhabit, remain in, and feed on flowers.
3. Sticky cards positioned near vents or doors where insects tend to enter greenhouses from outdoors may capture more adult insect pests, thus inflating the overall numbers on sticky cards.
4. The age structure of the insect pest population in which adults or immature stages are dominant at different times may underestimate numbers because the immature stages (e.g., nymphs or larvae) cannot be captured on colored sticky cards.
5. Sticky card color, whether yellow or blue, may affect sticky card counts. For instance, western flower thrips adults tend to be captured on blue sticky cards at lower densities than on yellow sticky cards.

6. The placement of sticky cards in regards to the crop canopy may impact the number of insects captured. Higher numbers of adult western flower thrips, for example, are captured on sticky cards located just above the crop canopy. Also, more fungus gnat adults will be captured on sticky cards that are placed near the growing medium surface.
7. The practice of removing plants or harvesting cut flowers can create disturbances within the greenhouse that agitates insect pests, causing them to be active and more dispersed within the greenhouse—resulting in higher sticky card counts.

An additional factor that can affect sticky card counts is the presence of horizontal air-flow fans (HAFs), which can spatially distribute insect pests throughout the greenhouse on air currents. In order to avoid problems with misleading sticky card counts, develop maps of greenhouses that include the location of each sticky card.

Action thresholds are likely to be more applicable in perennial cropping systems such as cut flowers than annual cropping systems (e.g., bedding plants), especially those with multiple horticultural crops and pests. Furthermore, action thresholds may not be feasible in greenhouse production systems due to the low tolerance for insect and mite pests. There is also the problem affiliated with attempting to correlate sticky card counts with actual insect populations present in the greenhouse.

In general, nearly all greenhouses will have some level of insect and/or mite pest activity. However, as long as the numbers of insect and mites pests are low enough, based on the action threshold, and are not causing problems, then it is difficult to justify the time, labor, and direct costs affiliated with applying pesticides. In fact, most greenhouse producers may spend 95% of their time attempting to suppress 5% of the pest population. Scouting is a very efficient method of assessing the numbers of insect and/or mite pests throughout the growing season and indirectly determining the effectiveness of pest management strategies. Subsequently, greenhouse producers may want to establish their own "realistic" action thresholds based on the current cropping systems in order to improve pest management decisions.

Chapter 4

Cultural Control and Sanitation

The implementation of cultural control and sanitation practices in greenhouses will help to alleviate potential problems with insect and mite pests. Both cultural (irrigation and fertility) and sanitation (weed removal) practices are the primary means of avoiding pest problems in greenhouse production systems.

Irrigation

Overwatering plants and the growing medium in which they reside exacerbates problems with fungus gnats and shore flies (described in Chapter 2, "Pest Identification") because both insect pests survive, develop, and reproduce more efficiently under excessively moist conditions associated with growing medium (depending on age and type). Also, excessively moist conditions promote the growth of algae (Figure 4.1), which provides a favorable substrate and breeding site for fungus gnats and shore flies. However, plants should not be underwatered as this causes stress and can increase

Figure 4.1   Algae underneath bench in greenhouse.

susceptibility to specific pests, such as the twospotted spider mite. Nonetheless, overhead irrigation will increase relative humidity, reducing development and reproduction of twospotted spider mite populations, and may also physically dislodge certain insect (e.g., aphids and thrips) and mite pests from plants.

Fertility

The misuse of fertilizers (e.g., overfertility) may change plant quality such that plants are a better food source for insect and mite pests, resulting in enhanced development, growth, and female reproduction. Overfertilizing plants, especially with nitrogen-based fertilizers, can stimulate new growth (Figure 4.2). New, young succulent leaves do not have a well-developed waxy layer (epidermis) and may have higher levels of amino acids (building blocks of proteins) that can lead to increased feeding by insect and mite pests with piercing–sucking mouthparts. Moreover, succulent growth

Figure 4.2 Succulent leaf growth of basil plants.

and an increased level of amino acids in plant tissues due to overfertilizing can enhance development and reproduction of certain insect and mite pests. Furthermore, overfertilized plants are much more susceptible to aphids, scales, leafminers, whiteflies, mealybugs, and spider mites.

Sanitation

Sanitation is the "first line of defense" in any pest management program, and can reduce potential problems with insect and mite pests. The effectiveness of pesticides or biological control (described in Chapters 6 and 7) is contingent on implementing a stringent sanitation program. Sanitation is one of the easiest and least expensive practices to implement because sanitation can be performed during normal operating hours. Sanitation involves the following: (1) removing weeds, (2) reducing algae, and (3) removing plant debris, from both within and outside the greenhouse facility.

Weeds inside and outside the greenhouse (Figures 4.3 through 4.5) provide refuge for insect and mite pests, such as aphids, leafminers (Figure 4.6), thrips, spider mites, and whiteflies, which allow these pests to survive and disperse to the main crop. Weeds that can serve as a refuge for

Figure 4.3   Weed (chickweed) growing underneath bench in gutter of greenhouse.

Figure 4.4   Weeds growing in soil underneath greenhouse bench.

Figure 4.5   Weeds growing outside of greenhouse opening.

Figure 4.6   Weed (oxalis) growing in plant container.

insects include sow thistle, *Sonchus* spp. (aphids and white-flies); oxalis, *Oxalis* spp. (thrips) (Figure 4.7); and dandelion, *Taraxacum officinale* (whiteflies). In addition, many weeds serve as reservoirs for pathogens, such as viruses that can be acquired by insects and then transmitted to the main crop during feeding. Weeds that serve as reservoirs for viruses, specifically the tospoviruses—*impatiens necrotic spot* and *tomato spotted wilt* viruses—include chickweed (*Stellaria media*), lambsquarters (*Chenopodium* spp.), nightshade (*Solanum* spp.), oxalis (*Oxalis* spp.), shepherd's purse (*Capsella bursa-pastoris*), pigweed (*Amaranthus* spp.), and bindweed (*Convolvulus* spp.).

One method of reducing problems with weeds is to install landscape or fabric barriers underneath benches (Figure 4.8), which are geotextile, nonbiodegradable materials that will prevent weeds from emerging from the soil underneath benches and also diminish algae growth. There are herbicides (weed killers) registered for use inside and outside greenhouses although caution must be exercised when using herbicides inside greenhouses. A preemergent

Figure 4.7 **Weed (pigweed) with leafminer larvae in leaves.**

Figure 4.8 **Weed fabric barrier.**

herbicide can be applied prior to weed emergence, whereas a postemergent herbicide can be applied after weeds emerge. However, be sure to avoid any inadvertent plant injury (such as phytotoxicity), only use herbicides, especially those that have systemic (when applied as a spray) and postemergent activity, by making applications when greenhouses are empty. Always read the label directions before mixing and loading. Large weeds (>15.2 cm [6 in.] in height) should be physically removed by hand, taking care to remove both the aboveground portion and roots. Weed-free zones or areas around the outside perimeter of greenhouses (3–9.1 m [10–30 ft.]) can reduce the *migration of insects*, such as winged adults of the western flower thrips, through openings, which decreases the potential incidence for disease transmission.

Algae provide an ideal breeding substrate for fungus gnats and shore flies; subsequently algae must be reduced or "eliminated" from benches and floors. The practices of not overwatering and overfertilizing plants, and using well-drained growing media will definitely help avoid problems with algae. Furthermore, reducing algae may be accomplished by using commercially available disinfectants such as those containing the following active ingredients: hydrogen peroxide, hydrogen dioxide, and/or quaternary ammonium chloride salts. Another practice is routinely pressure-washing growing medium from benches and walkways, which will alleviate algae buildup and consequently avoid having to deal with insect pests.

Plant debris, such as leaves and flowers, and growing medium debris (Figure 4.9) provide refuge for certain insect and/or mite pests. Insects and even mites can migrate to fresh plant material as plant debris desiccates. For example, reports indicate that plant material and growing medium debris placed into unsealed refuse containers (Figures 4.10 and 4.11) can be a source of insect pests. As plant material desiccates, adults can *migrate* to the main crop. Therefore, always

Figure 4.9 **Crop and growing medium debris in trash receptacle (dumpster).**

Figure 4.10 Debris placed into garbage container without lid.

Figure 4.11 **Plant and growing medium debris in garbage container without lid.**

place debris into refuse containers with tight-sealing lids (Figure 4.12). Also, any leftover growing medium provides sites for fungus gnat adults to lay eggs and western flower thrips to pupate. Use a broom or shop vacuum (Figure 4.13) to remove plant or growing medium debris. Moreover, old stock plants

Figure 4.12   Garbage container with tight-sealing lid.

Figure 4.13   Shop vacuum.

or those remaining at the end of the growing season should be removed because they can be a potential source of insect and mite pests. Old stock plants can also be a reservoir for the viruses transmitted by insects such as the western flower thrips. Another important sanitation practice is to immediately remove any plants heavily infested with insect or mite pests from the greenhouse.

Chapter 5

Physical Control

Screening or microscreening greenhouse openings, such as
vents and sidewalls, can inhibit the entry of winged insects
into greenhouses (Figure 5.1), which will reduce the number
of pesticide applications needed during the growing season.
The hole or pore size (mesh) of the screening (Figure 5.2)
will help decide the appropriate material to select, which
will be contingent on the insect pests to be excluded. Screen
size depends on the insect, whether it be aphids (340 µm
or 0.013 in.), leafminers (640 µm or 0.025 in.), thrips (192 µm
or 0.0075 in.), or whiteflies (462 µm or 0.018 in.). Select and
install insect screening based on the smallest insect pest to be
excluded. The installation of insect screening may decrease
the incidence of viral diseases, such as the *impatiens necrotic
spot* and *tomato spotted wilt* viruses that are vectored by
the western flower thrips, because the screening material,
depending on the mesh size, may exclude the insect pest
from entering the greenhouse. Furthermore, screening greenhouse vents can exclude larger insects such as aphids, beetles, and moths from entering the greenhouse, some of which
are capable of transmitting diseases. Screening will also
prevent weed seeds from entering the greenhouse, which will
alleviate potential problems with weed seeds establishing and

Figure 5.1   Insect screening located on outside of greenhouse vents.

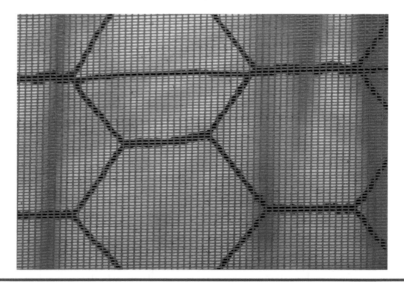

Figure 5.2   Close-up of insect screening mesh (pore size).

subsequently developing in flats or containers with growing medium (Figure 5.3).

Screen construction types include woven, knitted, and film. Some screening materials exclude insects that are smaller than the screen mesh size because of configuration,

Figure 5.3   **Weed growing in container with plant.**

pattern, and uniformity. However, not all screening materials are the same based on the insect pests excluded. The use of screening may reduce airflow through the greenhouse. Therefore, increasing the surface area through which air moves compensates for reduced airflow. Screens with smaller holes are more resistant to airflow. The National Greenhouse Manufacturers Association (NGMA) has information on the amount of screening needed in order to avoid restricting airflow and damaging fan motors (www.ngma. com). Large screened cages can be built to cover openings that allow for adequate airflow, and prevent fans from malfunctioning (Figure 5.4).

Not every greenhouse is amenable to screening. Screening can be cost prohibitive if the entire greenhouse has to be screened. Therefore, if screening the entire greenhouse is not an option, then concentrate on screening the windward sides, which are the sides that insect pests are most likely to enter. Screens must be cleaned periodically to remove debris that can reduce airflow and cooling system capacity. Always turn off fans before cleaning screens.

Figure 5.4   Installation of insect screening on south side of greenhouse.

Also, avoid using a high-pressure spray or brush, which can damage or create openings in the screening. Prior to selecting materials for screening greenhouse openings, greenhouse producers need to consider the cost of the material and installation, the economic value of the crop(s) grown, susceptibility of crop(s) to viral diseases, and insect pests targeted for exclusion. Furthermore, any costs should be compared to the potential savings obtained from reduced pesticide applications.

Expenses associated with screening depend on a number of factors including: (1) increase in surface area required to maintain adequate ventilation, (2) cost of screening material, (3) frequency of cleaning, and (4) number of openings that need to be screened. The surrounding area and subsequent crops being grown will dictate the types of insects that can migrate into the greenhouse through openings. Screening greenhouses will only be effective when used in conjunction with other pest management practices such as scouting.

Mass trapping has been suggested as a way to capture large numbers of insect pests, especially fungus gnat, leafminer, western flower thrips, and whitefly adults. This involves using yellow sticky tape that is positioned in rows hung vertically within the greenhouse (Figures 5.5 through 5.7). The use of yellow sticky tape may prove to be successful during propagation in substantially reducing numbers of fungus gnat and shore fly adults (Figure 5.8). Moreover, the placement of yellow sticky tape near openings may be helpful in capturing adult insects as they enter the greenhouse from outdoors (Figure 5.9). Another technique is to stretch the yellow sticky tape behind plants in flower, and then "blow" across the flowers using a gas-powered blower, to dislodge and capture western flower thrips on the yellow sticky tape. This technique may provide supplemental suppression of western flower thrips when used with other pest management strategies.

Figure 5.5   **Yellow sticky tape installed on greenhouse bench to capture fungus gnat and shore fly adults.**

Figure 5.6   **Yellow sticky tape used to mass-trap winged adult insects.**

Figure 5.7   **Yellow sticky tape installed in greenhouse to capture western flower thrips, leafminer, and whitefly adults.**

Figure 5.8   Close-up of yellow sticky tape showing captured insects.

Figure 5.9   Yellow sticky tape placed near opening to capture adult insects before they enter the greenhouse.

Chapter 6

Pesticides

Pesticides (in this case, insecticides and miticides) are an integral component of most greenhouse pest management programs. Pesticides, in general, are relatively inexpensive, easy to apply, and effective (in most cases). Additionally, there is a psychological satisfaction and level of comfort after applying pesticides to suppress insect and/or mite pest populations, which is different than using biological control (described in Chapter 7) where there may be level of uncertainty. Pesticides are used primarily to kill insect and mite pests, and maintain the aesthetic quality of greenhouse-grown horticultural crops for marketability and salability. Greenhouse producers use pesticides because, when selling whole plants, there is still a stigma associated with consumers that they will not purchase plants with either insects or mites, or plants that exhibit any noticeable damage symptoms. Furthermore, there are issues regarding the low tolerance of certain insects that vector diseases such as viruses, which can result in frequent use of pesticides. However, due to strict laws and regulations, and the extensive costs of registering a pesticide, fewer new active ingredients are being registered for use in greenhouses. Moreover, greenhouse-grown horticultural crops are considered a specialty crop, which is why many pesticides registered for use in greenhouses are substantially more expensive than those registered for use on agricultural crops, such as corn, soybean, cotton, and rice.

The types of pesticides that may be used in greenhouses are: (1) contact, (2) stomach poison, (3) translaminar, and (4) systemic. *Contact pesticides* kill an insect and/or mite pest by direct contact or when an insect and/or mite pest walks or crawls over a treated surface. The insect or mite pest walks across a treated surface and then pesticide residues enter the body and move to sites of action. The activity of a *stomach poison pesticide* is affiliated with an insect pest feeding on treated surfaces (e.g., leaves) and ingesting the pesticide residues, which are then absorbed through the stomach lining. Insects stop feeding in 24–48 hours with death usually occurring within 2–4 days. *Translaminar pesticides* work by penetrating leaf tissues and forming a reservoir of active ingredient within the leaf, which provides residual activity against foliar-feeding insects and mites. *Systemic pesticides* applied to the growing medium, as either a drench or granule, involve the active ingredient being taken up by the root system, which is then translocated or distributed throughout the plant. However, plants must have a well-established root system and be actively growing in order to take up the active ingredient. Systemic pesticides are used primarily to prevent infestations of phloem-feeding insects such as aphids, mealybugs, soft scales, whiteflies, and some leafhoppers. Always irrigate plants prior to applying any pesticide in order to maintain turgidity and thus avoid problems with plant injury (such as phytotoxicity).

The two major categories of pesticides that target insect and mite pests are *broad-spectrum* and *narrow-spectrum* (also referred to as selective) *pesticides*. *Broad-spectrum pesticides* are active on a multitude of different insect and mite pests, which is beneficial when there is a wide diversity of horticultural crops grown in greenhouses. *Narrow-spectrum pesticides* are only active on certain types of insect and/or mite pests.

Maximizing Pesticide Performance

In general, when pest suppression falters, pesticide resistance (described later) is initially blamed. Although this may be the case in some situations, there are a number of factors that may result in inadequate pesticide performance. These factors range from application techniques to water quality issues. Below are the primary factors associated with insufficient pesticide performance.

Pest Identification

Properly identify insect and mite pests prior to selecting a pesticide because many selective pesticides have a narrow range of target insect and/or mite pests that they are active against. For example, some pesticides are only registered for use on one group of pests (e.g., mites), whereas others may have activity on two or three different insect types (e.g., thrips and aphids). In order to correctly identify a given insect or mite pest, always have several reference publications available that contain clear images in order to help identify a pest in question. An alternative option is to send samples to a state extension entomologist, or university-based or independent plant diagnostic clinic. Once a given pest or pests has been correctly identified, then the appropriate pesticide can be applied. It is also important to understand the biology, behavior, and life cycle (refer to Chapter 2, "Pest Identification") of a given pest to determine where certain life stages are located on plants, and which life stage(s) is most susceptible to pesticides.

Coverage

Thorough, uniform coverage of all aboveground plant parts including leaves, stems, and flowers is important in suppressing insect and mite pest populations with pesticides, especially pesticides with contact activity. Both the upper and lower leaf surfaces must receive sufficient volume of the spray solution. Determine the location of pests, and then direct spray applications to those plant parts to obtain maximum coverage and thus increase pesticide effectiveness. Since most pesticides have contact activity, leaf undersides must be sufficiently covered (Figures 6.1 and 6.2), which is where the majority of life stages (eggs, larvae/nymphs, pupae, and adults) of certain pests such as the twospotted spider mite, greenhouse whitefly, and sweet potato whitefly are located. However, some pesticides have translaminar or localized activity, which means the pesticide solution or residues penetrate leaf tissues and form a reservoir of active ingredient within the leaf. Pesticides with translaminar properties provide supplemental residual activity against insect and mite pests with piercing–sucking mouthparts even after residues dry. The effective residual activity may be 14–40 days; however, residual activity depends on the specific pesticide and plant type. The use of water sensitive paper or spray cards can be used to assess spray coverage

Figure 6.1   Foliar application of pesticide in greenhouse.

Figure 6.2 **Foliar application of pesticide (note that applicator is targetting spray at the leaf underside).**

by quantifying spray droplet distribution and deposition. The strips of water-sensitive paper turn blue when exposed to water droplets (Figures 6.3 and 6.4). Spray cards can be randomly distributed among a crop and securely attached to plants. This practice should be conducted routinely to determine droplet size and density. Furthermore, using spray cards will help to evaluate the performance of the spray equipment and efficiency of the applicator.

Conduct spray applications when the person performing the application is not too tired in order to avoid insufficient spray coverage. Pesticide applications should never be performed during the heat of the day when discomfort can lead to reduced spray coverage and plant injury (such as phytotoxicity) (Figures 6.5 and 6.6). In addition, applications made during the heat of the day may possibly result in the applicator experiencing heat exhaustion due to the personal protective clothing and equipment required.

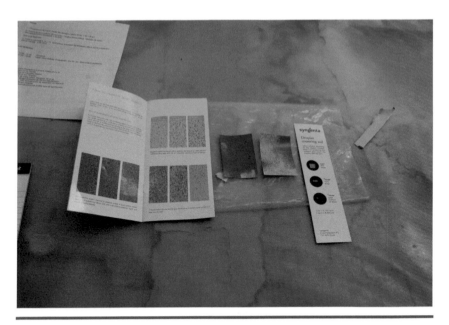

Figure 6.3 Water-sensitive paper used to determine spray coverage by applicator.

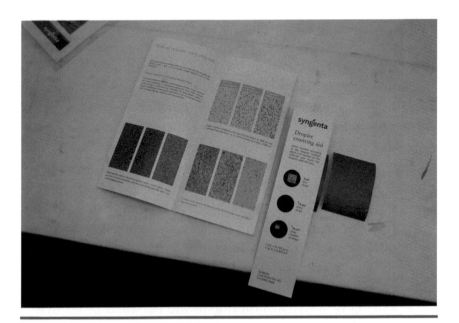

Figure 6.4 Water-sensitive paper with examples of spray droplet patterns.

Figure 6.5   Injury (such as phytotoxicity) to marigold leaf caused by pesticide application.

Figure 6.6   Injury (such as phytotoxicity) to veronica plant caused by pesticide application.

Timing of Application

Pesticide applications conducted when insect or mite pest populations are extensive results in taking longer to lower the numbers below damaging levels. Therefore, more frequent applications will be required, especially when dealing with multiple age structures or overlapping generations. Furthermore, insect and/or mite pests may: (1) have developed into life stages, such as eggs and pupae, that are tolerant of pesticide applications; (2) already be causing substantial plant damage; or (3) be in locations on plants, such as unopened flower buds, that are difficult to reach with sprays. Always time pesticide applications when pest numbers are low, which is based on information obtained from scouting (refer to Chapter 3, "Scouting").

Pesticides should be applied either in the early morning or late afternoon because this is generally when most insect and mite pests are active (although this is contingent on ambient air temperature). If pesticides are applied when insect and mite pests are minimally active (because temperatures are too high) then less efficacy (based on mortality) may result, particularly when using contact pesticides. Pesticides applied during hot, dry sunny days can lead to rapid drying and thus less residual activity, which will reduce overall effectiveness. Horticultural oils applied during overcast days may result in plant injury (such as phytotoxicity) because the material does not dry, whereas applying pesticides during evening hours is more amenable, especially during the summer. However, pesticide applications conducted in the evening may promote disease development such as gray mold (*Botrytis cinerea*) due to the extended period of leaf wetness. Nonetheless, drying through heating, venting, and/or the use of horizontal air-flow fans can alleviate problems with foliar diseases.

Water Quality

Issues associated with water quality relating to spray solution pH may influence pesticide effectiveness. The potential
hydrogen (pH) refers to a scale ranging from 1 to 14. A pH
of 7 is neutral, whereas a pH above 7 is alkaline (basic) and
below 7 is acidic. The pH is a logarithmic scale and the sensitivity of a pesticide to water pH will increase by a factor of
10 for every pH unit. For example, a pH of 6 is 10 times
more acidic than a pH of 7, and a pH of 5 is 100 times more
acidic than a pH of 7. A spray solution pH greater than 7 may
result in certain pesticides fragmenting due to *alkaline hydrolysis*, in which a pH greater than 7 causes chemical degradation of certain pesticides. Alkaline water fragments pesticide
molecules resulting in release of individual ions (electrically
charged atoms) that can reassemble with other ions. These
new combinations do not have any insecticidal or miticidal
activity, thus reducing the effectiveness of a pesticide spray
application. Some pesticides, however, can undergo acid
hydrolysis at a pH less than 7. The rate of hydrolysis depends
on: (1) the pH of the water and spray solution, (2) pesticide
chemical properties, (3) the length of time the spray solution
resides in the spray container, and (4) the water temperature in the spray container. Moreover, the exposure time in
the alkaline spray solution is important. For instance, the
spray emitted from the nozzle during the first hour of a foliar
application may be more effective than what is emitted the
last hour of a foliar application. A 10°C (18°F) increase in the
spray solution temperature may double the rate of decompo-
sition. For example, at a pH of 9 and water temperature of
25°C (77°C), acephate (Orthene) loses 50% of activity in 1–2
days. Doubling the temperature will expedite the rate of degradation twofold. Exposure of the spray container to direct
sunlight can also impact the rate of hydrolysis.

Pesticide manufacturers have information associated with the effect of water pH on the half-life of a pesticide. Half-life is the time required for 50% of an active ingredient to hydrolyze or break down, or the length of time that a pesticide's original strength is diminished by 50%. Insecticides, in general, are more susceptible to *alkaline hydrolysis* than fungicides or even plant growth regulators. Insecticides in the chemical classes organophosphate (acephate and chlorpyrifos), carbamate (methiocarb), and pyrethroid (bifenthrin, cyfluthrin, fenpropathrin, fluvalinate, and lambda-cyhalothrin) are highly sensitive to *alkaline hydrolysis* when the pH of the spray solution is greater than 7. In fact, carbamates will degrade faster than organophosphates. Therefore, always monitor spray solution pH and adjust accordingly in order to maintain pesticide effectiveness. The appropriate pH range for most pesticides is between 5 and 7; however, a number of pesticides perform better at or above a pH of 7. A water pH above 7 can also increase the length of time required to dissolve pesticides formulated as water-soluble packets.

Water pH can be adjusted, although the process should be performed carefully. The use of pH paper is not an accurate means of monitoring pH within 0.5; however, a water pH between 6 and 7 may be acceptable such that the use of pH paper may be valid. Acetic acid (vinegar) can be added to the spray solution to lower the pH below 7 using small increments in conjunction with periodically checking the spray solution with pH paper. Be sure to avoid adding too much vinegar so the spray solution pH is maintained around 6.5. If the spray solution pH needs to be increased, then add household ammonia. Always adjust the water pH before adding any pesticides to the spray container. Water pH can also be adjusted by using buffering or water-conditioning agents (adjuvants). These adjuvants reduce the potential for *alkaline hydrolysis* by modifying the spray solution pH so it is easier to maintain within a range of 5–7. These adjuvants are typically safer in lowering the spray solution pH compared to materials such as sulfuric acid. However, always add adjuvants to the spray container prior to adding pesticides as certain pesticides may degrade initially when dissolved in an alkaline solution. There are ways to avoid pH issues including: (1) checking the pesticide label for precautions associated with using high pH water sources, (2) regularly monitoring water pH, and (3) not leaving pesticide solutions in spray containers for extended periods of time.

Rotating Pesticides with Different Modes of Action

The rotation of pesticides with different modes of action will help to avoid the prospect of insect and mite pest populations developing resistance (refer to the section "Understanding Issues Associated with Pesticide Resistance"). Failure to rotate modes of activity—not chemical class—can result in resistance and thus reduced insect and/or mite pest population suppression. Always use the same mode of action within an insect or mite pest generation before switching to a different mode of action (refer to Tables 6.1 and 6.2).

Application Technique and Formulation

Aerosols and fine sprays are effective against winged adults, whereas high-volume spray applications are preferred for use against sedentary or immobile life stages and pests inhabiting the growing medium, such as the larval stage of fungus gnats. An aerosol used against insect and/or mite pests located in the growing medium, underneath leaves, or deep within the crop canopy will result in inadequate suppression because of insufficient coverage. Aerosols and fumigants should be used against adults, whereas drenches can be used against the larval stages of pests located in the growing medium, such as fungus gnats and shore flies, because the application is targeting the susceptible life stage (e.g., larvae). Low-volume applications are more effective when plants are small, but may not

Table 6.1 Mode of Action and Pest Activity Chart

Mode of Action	Pest Control Materials	IRAC Group	Type	WF	APH	THRIPS	MB	SM	FG	SF	LM	CAT
Acetylcholineesterase Inhibitors	Acephate (Orthene)	1B	C, S, T	X	X	X	X					
	Chlorpyrifos (DuraGuard)	1B	C		X	X	X		X	X	X	X
	Methiocarb (Mesurol)	1A	C		X	X						
Prolong opening of sodium channels	Bifenthrin (Talstar/Attain)	3A	C	X	X	X	X	X	X			X
	Cyfluthrin (Decathlon)	3A	C	X	X	X	X		X			X
	Fenpropathrin (Tame)	3A	C	X	X	X	X	X			X	X
	Fluvalinate (Mavrik)	3A	C	X	X	X		X				X

(Continued)

Table 6.1 (Continued)   Mode of Action and Pest Activity Chart

Mode of Action	Pest Control Materials	IRAC Group	Type	Pest Activity (Based on Label)								
				WF	APH	THRIPS	MB	SM	FG	SF	LM	CAT
Prolong opening of sodium channels	Lambda-cyhalothrin (Scimitar)	3A	C	X	X	X	X	X			X	X
Nicotinic acetylcholine receptor disruptors	Acetamiprid (TriStar)	4A	C, S, T	X	X	X	X		X		X	X
	Dinotefuran (Safari)	4A	C, S, T	X	X	X	X		X		X	
	Imidacloprid (Marathon)	4A	C, S, T	X	X	X	X		X		X	
	Thiamethoxam (Flagship)	4A	C, S, T	X	X		X		X			

(Continued)

Table 6.1 (Continued) Mode of Action and Pest Activity Chart

Mode of Action	Pest Control Materials	IRAC Group	Type	Pest Activity (Based on Label)								
				WF	APH	THRIPS	MB	SM	FG	SF	LM	CAT
Nicotinic acetylcholine receptor agonist and GABA chloride channel activator	Spinosad (Conserve)	5	C, ST, T			X		X			X	X
GABA chloride channel activator	Abamectin (Avid)	6	C, T	X	X	X		X			X	
Juvenile hormone mimics	Fenoxycarb (Preclude)	7B	C	X	X	X	X	X			X	X
	Kinoprene (Enstar)	7A	C	X	X	X	X		X			
	Pyriproxyfen (Distance/Fulcrum)	7C	C, T	X	X		X		X	X		
Chitin synthesis inhibitors	Buprofezin (Talus)	16	C	X			X					

(Continued)

Table 6.1 (Continued) Mode of Action and Pest Activity Chart

Mode of Action	Pest Control Materials	IRAC Group	Type	Pest Activity (Based on Label)								
				WF	APH	THRIPS	MB	SM	FG	SF	LM	CAT
Chitin synthesis inhibitors	Cyromazine (Citation)	17	C						X	X	X	
	Diflubenzuron (Adept)	15	C	X					X	X	X	X
	Etoxazole (TetraSan)	10B	C, T					X				
	Novaluron (Pedestal)	15	C	X		X					X	X
Ecdysone antagonist[a]	Azadirachtin (Azatin/Ornazin/Molt-X/AzaGuard)		C, ST	X	X	X	X		X	X	X	X
Ecdysone agonist[b]	Methoxyfenozide (Intreprid)	18	ST									X

(Continued)

Table 6.1 (Continued)   Mode of Action and Pest Activity Chart

Mode of Action	Pest Control Materials	IRAC Group	Type	Pest Activity (Based on Label)								
				WF	APH	THRIPS	MB	SM	FG	SF	LM	CAT
Growth and embryogenesis inhibitors	Clofentezine (Ovation)	10A	C					X				
	Hexythiazox (Hexygon)	10A	C					X				
Selective feeding blockers	Flonicamid (Aria)[c]	9C	C, S, T	X	X	X	X					
	Pymetrozine (Endeavor)	9B	C, S, T	X	X							
Disruptors of insect midgut membranes	*Bacillus thuringiensis* subsp. *israelensis* (Gnatrol)	11	ST						X			

(Continued)

Table 6.1 (*Continued*)   Mode of Action and Pest Activity Chart

Mode of Action	Pest Control Materials	IRAC Group	Type	Pest Activity (Based on Label)								
				WF	APH	THRIPS	MB	SM	FG	SF	LM	CAT
Disruptors of insect midgut membranes	*Bacillus thuringiensis* subsp. *kurstaki* (Dipel)	11	ST									X
Oxidative phosphorylation uncoupler	Chlorfenapyr (Pylon)	13	C, T			X		X	X			X
Oxidative phosphorylation inhibitor	Fenbutatin oxide (ProMite)	12B	C					X				
Mitochondria electron transport inhibitors	Acequinocyl (Shuttle)	20B	C					X				
	Bifenazate (Floramite)		C					X				
	Cyflumetofen (Sultan)	25	C					X				

(*Continued*)

Table 6.1 (Continued)   Mode of Action and Pest Activity Chart

Mode of Action	Pest Control Materials	IRAC Group	Type	Pest Activity (Based on Label)									
				WF	APH	THRIPS	MB	SM	FG	SF	LM	CAT
Mitochondria electron transport inhibitors	Fenazaquin (Magus)	21A	C	X				X				
	Fenpyroximate (Akari)	21A	C				X	X				
	Pyridaben (Sanmite)	21A	C	X				X				
	Tolfenpyrad (Hachi-Hachi)	21A	C	X	X	X						X
Selective activation of ryanodine receptors	Cyantraniliprole (Mainspring)	28	C, ST, T	X	X	X					X	X
Desiccation or membrane disruptors	Neem oil (Triact)		C	X	X	X	X	X				
	Mineral oil (Ultra-Pure Oil/ SuffOil-X)		C	X	X	X	X	X	X		X	

(Continued)

Table 6.1 (Continued) Mode of Action and Pest Activity Chart

Mode of Action	Pest Control Materials	IRAC Group	Type	Pest Activity (Based on Label)								
				WF	APH	THRIPS	MB	SM	FG	SF	LM	CAT
Desiccation or membrane disruptors	Potassium salts of fatty acids (M-Pede)		C	X	X	X	X	X				
Lipid biosynthesis inhibitor	Spiromesifen (Judo)	23	C, T	X				X				
	Spirotetramat (Kontos)	23	C, T, S	X	X		X	X				
Nicotinic acetylcholine receptor agonist and GABA chloride channel activator + nicotinic acetylcholine receptor disruptor	Spinetoram + Sulfoxaflor (XXpire)	5 + 4C	C, T, S	X	X	X	X					X

(Continued)

Table 6.1 (Continued) **Mode of Action and Pest Activity Chart**

Mode of Action	Pest Control Materials	IRAC Group	Type	Pest Activity (Based on Label)								
				WF	APH	THRIPS	MB	SM	FG	SF	LM	CAT
Unknown	Beauveria bassiana (BotaniGard/Naturalis)		C	X	X	X	X					X
	Isaria fumosoroseus (NoFly/Preferal)		C	X	X	X	X	X	X		X	
	Metarhizium anisopliae (Met52)		C	X		X		X				
	Pyridalyl (Overture)		C, ST, T			X						X
	Pyrifluquinazon (Rycar)		C, ST, T	X	X	X	X					

a Antagonist: A substance that acts against and blocks an action.

b Agonist: A substance acting like another substance that stimulates an action.

c Also blocks action of potassium channels.

Pest Activity Codes: WF—Whiteflies; APH—Aphids; MB—Mealybugs; SM—Spider mites; FG—Fungus gnats; SF—Shore flies; LM—Leafminers; CAT—Caterpillars. Type Codes: C—Contact; S—Systemic; T—Translaminar; ST—Ingested. IRAC—Insecticide Resistance Action Committee.

Table 6.2 Miticides (Active Ingredient and Trade Name) and Twospotted Spider Mite Life Stages, and Mode of Action Chart

Active Ingredient	Trade Name	IRAC Group	Activity Type	Egg	Larva	Nymph	Adult	Mode of Action
Abamectin	Avid	6	C and T		X	X	X	GABA[a] chloride channel activator
Abamectin + Bifenazate	Sirocco	6	C and T	X	X	X	X	GABA chloride channel activator + mitochondria electron transport inhibitor
Acequinocyl	Shuttle	20B	C	X	X	X	X	Mitochondria electron transport inhibitor
Bifenazate	Floramite		C	X	X	X	X	Mitochondria electron transport inhibitor
Chlorfenapyr	Pylon	13	C and T		X	X	X	Oxidative phosphorylation uncoupler

(Continued)

Table 6.2 (*Continued*) Miticides (Active Ingredient and Trade Name) and Twospotted Spider Mite Life Stages, and Mode of Action Chart

Active Ingredient	Trade Name	IRAC Group	Activity Type	Egg	Larva	Nymph	Adult	Mode of Action
Clofentezine	Ovation	10A	C	X	X	X		Growth and embryogenesis inhibitor
Cyflumetofen	Sultan	25	C	X	X	X	X	Mitochondria electron transport inhibitor
Etoxazole	TetraSan	10B	C and T	X	X	X		Chitin synthesis inhibitor
Fenazaquin	Magus	21A	C	X	X	X	X	Mitochondria electron transport inhibitor
Fenbutatin oxide	ProMite	12B	C		X	X	X	Oxidative phosphorylation inhibitor
Fenpyroximate	Akari	21A	C	X	X	X	X	Mitochondria electron transport inhibitor

(Continued)

Table 6.2 (*Continued*) Miticides (Active Ingredient and Trade Name) and Twospotted Spider Mite Life Stages, and Mode of Action Chart

Active Ingredient	Trade Name	IRAC Group	Activity Type	Egg	Larva	Nymph	Adult	Mode of Action
Hexythiazox	Hexygon	10A	C	X	X	X		Growth and embryogenesis inhibitor
Pyridaben	Sanmite	21A	C	X	X	X	X	Mitochondria electron transport inhibitor
Spiromesifen	Judo	23	C and T	X	X	X		Lipid biosynthesis inhibitor
Spirotetramat	Kontos	23	C, T, and S	X	X	X		Lipid biosynthesis inhibitor

Note: Additional materials that may be used include those with mineral oil, neem oil, or potassium salts of fatty acids as the active ingredient.

[a] GABA—Gamma-aminobutyric acid.

Activity Type Codes: C—Contact; T—Translaminar; S—Systemic.

IRAC—Insecticide Resistance Action Committee.

provide sufficient coverage when the crop canopy is dense, which may result in the need to use high-volume applications.

Target Pest Life Stage

Failure to suppress pest populations with pesticides can occur when the most susceptible life stage or stages of target insect and/or mite pests are not predominantly present. In general, the immature or young stages (larva or nymph) and adult are more susceptible to pesticides than the egg and pupal stage. In fact, most contact and systemic pesticides have no activity on the egg and pupa of many insect pests. For example, sufficient suppression will not occur if western flower thrips eggs and pupae are the predominant life stages present during an application. Young larvae that hatch from eggs and adults that emerge from pupae will have escaped pesticide exposure for several days following an application, which is especially critical with short residual pesticides, consequently resulting in the need for additional spray applications. Targeting the early development stages may reduce the number of pesticide applications required thus decreasing selection pressure on insect and mite pest populations. Proper scouting can help detect the presence of susceptible life stages (larvae, nymphs, and adults) and then pesticides can be applied accordingly, thus maximizing effectiveness. As indicated previously (refer to Chapter 2, "Pest Identification"), understanding the pest biology, behavior, and life cycle will help to determine what life stages are most susceptible to pesticides.

Label Rate

Following recommended label rates will maximize the effectiveness of pesticide applications. Exceeding the label rate may result in plant injury (such as phytotoxicity) to a crop leading to an economic loss, whereas using less than the recommended label rate may result in insufficient pest suppression. Abiding by the label rate offers the greatest potential for success in suppressing pest populations. When a range of rates are provided on a label (e.g., 6–12 fl. oz./100 gal.) use the low (6 fl. oz./100 gal.) or middle (10 fl. oz./100/gal.) label rates initially. Always avoid constantly using the highest label rate of a pesticide because this may result in problems when the highest label rate fails to provide sufficient suppression of insect and/or mite pest populations. Furthermore, using the highest label rate may increase the selection pressure on pest populations and consequently result in resistance developing faster. In fact, the lowest label rate may be just as effective as the highest label rate especially when initially making pesticide applications early in the crop production cycle.

Shelf Life

Pesticides are not designed to last indefinitely. Pesticides must be used within a specified time period—approximately 3–5 years depending on formulation—and before the expiration date. Many pesticides degrade when exposed to continuous extremes of hot (>37°C or 100°F) and cold (<0°C or 32°F) over an extended period of time, which then reduces pesticide effectiveness. Liquid formulations, if not used within 4 years, may eventually separate or precipitate out of solution, forming layers at the bottom of containers, which makes it difficult to resuspend the active ingredient into the solution so that the pesticide is suitable for use. Proper storage will help preserve pesticide shelf life. Insulated pesticide storage chambers are ideal for protecting pesticides from extreme environmental conditions, including temperature and sunlight (Figure 6.7). In general, the proper storage conditions for most pesticides are temperatures between 15°C and 21°C (60°F and 70°F) and a relative humidity between 40% and 60%.

Figure 6.7 **Pesticide storage container.**

Frequency of Application

Most pesticides only kill the young (larvae or nymphs) and adult stages of insect and mite pests with no direct effect on the eggs and pupae. Therefore, repeat applications are necessary to kill stages that were initially missed by previous applications such as larvae or nymphs that were in the egg stage, and adults that were in the pupal stage. When dealing with overlapping generations and different age structures simultaneously, repeat pesticide applications are required. In some cases, depending on the target insect and mite pest, 2–3 applications may be needed when pest populations are abundant. Moreover, frequency of application depends on the season. For instance, during cooler temperatures, the insect and mite life cycle (egg to adult) and the length of development between generations may be extended compared to warmer temperatures, which can result in fewer pesticide applications. A common problem is that intervals between spray applications are too long (e.g., >14 days) resulting in insufficient suppression of pest populations.

Understanding Issues Associated with Pesticide Resistance

Resistance is an inherited trait and evolution of resistance in insect or mite pest populations depends on existing genetic variability that permits survival of some individuals when exposed to a pesticide. Surviving individuals can then transfer traits (genetically) to the next generation, thus enriching the gene pool with resistant genes. The selection pressure or proportion of the pest population killed by a pesticide is a primary factor, along with genetic variation in the insect or mite pest population that is responsible for susceptibility to the pesticide, which influences the development of resistance. Whenever a pest population is exposed to a pesticide, the result is selection for resistance, which increases the frequency or proportion of resistant genes within an insect or mite pest population. Furthermore, resistance is an international problem, with expanding global trade of plant material that not only can disperse insect and mite pests, but may also spread resistant genes that pests harbor.

The speed at which resistance develops in an insect or mite pest population is dependent mainly on two biological parameters: (1) *short development time* and (2) *high female reproductive capacity.* Also, some pests, including the twospotted spider mite and the western flower thrips, possess a haplo-diploid breeding system in which males only have one set of chromosomes, so any new genetic features, arising from mutations, will be immediately expressed or fixed. A haplo-diploid breeding system increases the rate in which resistance can develop. Furthermore, genes associated with resistance are fully expressed in haploid (single set of chromosomes) males in haplo-diploid species, whereas with entirely diploid (double set of chromosomes) species, resistance may be partially hidden as recessive or codominant traits.

An individual does not become resistant, but due to frequent applications of a given pesticide over multiple generations, susceptible individuals are removed from the population and resistant individuals survive to breed and reproduce. Consequently, insect or mite pest populations are no longer suppressed by a given pesticide. Resistance can also develop due to the movement of insect or mite pests within and into greenhouses. There are three ways that pest immigration enhances resistance. First, migration from other horticultural crops within the greenhouse, or between greenhouses, increases the likelihood that pest populations have already been exposed to previous pesticide applications. Second, receiving plants from a distributor with insect or mite pests that were previously exposed to pesticides may enhance the development of resistance because most of these pests may already possess genes for resistance. Third, insect or mite pests that enter greenhouses from field or vegetable-grown crops grown outdoors may have been subsequently exposed to agricultural pesticides that are similar (in regards to mode of action) to those commercially available for use in greenhouse production systems.

Different mechanisms may confer resistance in various pest populations of the same species, and multiple resistance mechanisms may coexist in certain insect and mite pest populations. The mechanisms associated with resistance are: (1) metabolic, (2) physiological, (3) physical, (4) behavioral, and (5) natural. *Metabolic resistance* is degradation of the active ingredient by the pest. When a pesticide enters the body, enzymes attack and convert and/or detoxify the active ingredient into a nontoxic form. A number of enzymes may be involved in the detoxification process, including hydrolases (carboxylesterases), glutathione *S*-transferases, and cytochrome P450 mono-oxygenases or mixed function oxidases. These detoxifying enzymes convert pesticides (which are hydrophobic or "water hating") to be more "water loving" (hydrophilic) and less biologically active compounds that are eliminated via excretion. *Physiological resistance*, also referred to as target site insensitivity, is associated with the interaction between the pesticide and the designated target, which is similar to a key (pesticide) fitting into a lock (target site). Decreased binding affiliated with physiological resistance is analogous to the lock having been changed so that the key does not fit, and consequently the pesticide is no longer effective. Examples of this type of resistance occur in the organophosphate, carbamate, and pyrethroid chemical classes. Insect and mite pests can evolve different means to decrease susceptibility to organophosphate and carbamate pesticides including: (1) reduced sensitivity of acetylcholinesterase (ACHE), which is an enzyme in the central nervous system that is inhibited by pesticides in the organophosphate and carbamate chemical classes; (2) increased activity of ACHE; and/or (3) overproduction of ACHE. Furthermore, certain insect pests may possess what is known as knockdown resistance (kdr) where the nervous system has reduced sensitivity to pesticides in the pyrethroid chemical class due to modifications in the sodium channels associated with nerve axons. The nerve axons are the target site for pyrethroid pesticides. *Physical resistance* is a modification or alteration in the insect skin (cuticle) that reduces or delays penetration of the pesticide. Any delayed penetration through the cuticle diminishes the concentration of the pesticide active ingredient at the target site, thus preventing the possibility of overloading a pest's detoxification system. *Behavioral resistance* occurs when insect or mite pests avoid contact with a pesticide, which is associated with pests hiding in locations such as terminal growing points that are difficult for the pesticide to penetrate. *Natural resistance* is a general type of resistance pertaining to the lack of susceptibility of a pest population to a pesticide that is preexisting and is not affiliated with repeated exposure. Natural resistance may be due to any of the previously described metabolic, physiological, physical, or behavioral traits, and certain life stages not being susceptible to a given pesticide. For example, most pesticides are not directly effective against the egg and pupa life stages.

Furthermore, it is important to understand the differences between cross and multiple resistances in order to comprehend the complexity of resistance. *Cross resistance* is based on a single mechanism conferring resistance to pesticides in the same chemical class and/or having similar modes of action. *Multiple resistance* occurs in insect or mite pest populations that are resistant to more than one pesticide by means of more than one mechanism.

Factors that can affect the rate of resistance development in pest populations are divided into operational factors, which are under the direct control of greenhouse producers, and biological factors that are intrinsic to the pest population. Both operational and biological factors are presented in the following.

1. Operational Factors

a. Length of exposure to a single pesticide (characteristics associated with pesticide residues)

b. Frequency of pesticide applications

c. Rate of pesticide applied

d. Spray coverage (nonuniform deposition of spray droplets on leaves)

e. Level of mortality (proportion or percentage of pest population killed)

f. Applying pesticides when the most susceptible life stage or life stages such as the larva, nymph, and adult are not predominant in the pest population

g. Previous history of pesticides used—especially from suppliers

h. Relatedness of a pesticide to those that have been applied previously

i. Presence or absence of refuge sites or hiding places for pest populations

2. Biological Factors

a. Time required to complete one life cycle (egg to adult) or generation
b. Numbers of offspring (young) produced per generation
c. Mobility of individuals in pest population, which may be associated with winged adults dispersing to mate and/or feeding in protected habitats, which influences exposure to pesticides
d. Pests feed on a wide range of host plants, which may then allow preadaptation of pests to detoxify pesticides
e. Genetic system including parthenogenesis, haplo-diploid, or sexual reproduction
f. Expression of a resistant trait or traits due to either monogenic (when only one gene confers resistance resulting in rapid development of resistance) or poly-genic (when more than one gene confers resistance leading to slow resistance development) characteristics

The rate of resistance development in an insect or mite pest population may be increased due to greenhouse conditions. For example, environmental parameters such as temperature, relative humidity, and light intensity are typically conducive for rapid insect and mite pest development and reproduction. A greenhouse generally encloses insect and mite pests thus restricting susceptible individuals from migrating into the population. Therefore, resistant individuals within an insect or mite pest population are dominant can breed continuously in the greenhouse. However, susceptible individuals from areas outside the greenhouse not exposed to pesticides may not be able to enter and breed with resistant individuals. Furthermore, natural enemies (or biological control agents) such as parasitoids and predators are often absent, or present at low numbers, and also may not be able to immigrate into greenhouses. Finally, intensive year-round production in many greenhouses provides a continuous food supply for insect and mite pests resulting in multiple generations during the production cycle and frequent exposure to pesticide applications.

Resistance management is a strategy designed to sustain the effectiveness of currently existing pesticides, which primarily involves judicious selection and appropriate application of pesticides, and integration with other pest management strategies consistent with the basic pest management philosophy. Overall, resistance management is the most effective means to avoid or mitigate resistance occurring in insect and mite pest populations. Since resistance is genetically based, the frequency of resistance in a pest population is being mitigated in a resistance management program. Below are generalized guidelines to minimize the prospect of insect and mite pest populations developing resistance to a given pesticide:

1. Scout crops regularly to determine population dynamics of pest populations, and help time pesticide applications to target the most susceptible life stages (larvae, nymphs, and adults).
2. Implement proper cultural (irrigation and fertility) and sanitation (weed and plant debris removal) practices.
3. Install insect screening on greenhouse openings to prevent migrations of winged insect pests into greenhouses.
4. Implement the use of natural enemies (or biological control agents).
5. Use synergists when applying pesticides to inhibit enzymes involved in detoxification. However, be sure to read the pesticide label to determine if a synergist is already incorporated into the formulation.
6. Rotate pesticides with different modes of action within a generation of a target insect and/or mite pest.
7. Use pesticides with broad modes of activity such as insect growth regulators, insecticidal soap (potassium salts of

fatty acids), horticultural oils (mineral and neem based), selective feeding blockers, beneficial fungi and bacteria, and microorganisms (spinosad).

When developing a resistance management program always rotate common names (active ingredients)—not trade or brand names. Moreover, in order to alleviate the possibility of a pest population developing resistance, be sure to design rotation programs that involve using pesticides with discrete modes of action. Mode of action or mode of activity refers to how a pesticide affects the metabolic or physiological processes in an insect or mite pest. Previously, the idea was to rotate pesticides based on using different chemical classes; however, some chemical classes have similar modes of activity. For instance, organophosphates and carbamates, despite being different chemical classes, both have the same mode of action. These chemical classes inhibit the action of ACHE, an enzyme that deactivates the neurotransmitter acetylcholine thus halting nerve signals, which results in the total loss of nerve functions. For example, applying acephate (Orthene) consecutively within a generation and then switching to methiocarb (Mesurol) is not a proper rotation program because both pesticides have analogous modes of action. Similarly, although acequinocyl (Shuttle O), pyridaben (Sanmite), fenpyroximate (Akari), bifenazate (Floramite), tolfenpyrad (Hachi Hachi), cyflumetofen (Sultan), and fenazaquin (Magus) are in different chemical classes, all are active on the mitochondria electron transport system (responsible for energy production). The pesticides mentioned above work by inhibiting nicotinamide adenine dinucleotide hydride (NADH) dehydrogenase (complex I), binding to the Q_o center of cytochrome bc_1 (complex II), or acting on the NADH-C_oQ reductase site (complex III) in the mitochondria electron transport chain resulting in blocking synthesis of adenosine triphosphate (ATP). Therefore, pesticides that are active on the mitochondria electron transport system should not be used in succession. The neonicotinoid chemical class contains a number of pesticides including imidacloprid, thiamethoxam, acetamiprid, and dinotefuran. All neonicotinoid pesticides have similar modes of action, which involves binding to the postsynaptic nicotinergic ACh receptors, and consequently causing irreversible blockage and loss of nerve functions. As such, avoid using these pesticides in succession as this may increase selection pressure on exposed pest populations, possibly leading to resistance. The mode of action of all pesticides (insecticides and miticides), and then specifically miticides, registered for use in greenhouse production systems are presented in Tables 6.1 and 6.2.

A recommended strategy is to rotate pesticides with specific modes of action (primarily active on the central nervous system) with those having nonspecific, multiple, or broad modes of activity such as insect growth regulators, insecticidal soaps, horticultural oils, selective feeding blockers, beneficial fungi and bacteria, and microorganisms. The use of pesticides with broad modes of activity should minimize the rate at which insect or mite pest populations develop resistance. However, it is also important to rotate insect growth regulators with different modes of action since certain insect pests, including aphids and whiteflies, have exhibited resistance to a number of insect growth regulators.

In general, the rotation of different modes of action should occur every 2–3 weeks, or within one to two pest population generations, although the rotation of pesticides will depend on the time of year as temperature influences the duration of the life cycle (egg to adult). High temperatures, for example, that typically occur from midspring through summer (although this is dependent on geographic location) may shorten the development time of insect and mite pests, which oftentimes leads to the simultaneous presence of overlapping generations with variable age structures (eggs, larvae, nymphs, pupae, and adults). In addition to scouting more often, more frequent pesticide applications will be required along with continually rotating different modes of action. In contrast, during winter

months, the development time of insect and mite pest populations will be extended due to cooler temperatures and shorter
daylengths, thus decreasing the number of pesticide applications and modes of action required in rotation programs.
Furthermore, mixing or combining pesticides with different modes of action may delay resistance developing within
insect and/or mite pest populations because the mechanisms
required to resist mixtures may not be widespread or exist in
pest populations. Moreover, individuals in the pest population
may have difficulty developing resistance to several modes of
action simultaneously. Pesticide mixtures will be effective only
if insect and/or mite pests in the population that are resistant
to one pesticide succumb to the other pesticide in the mixture. However, pesticide mixtures may result in selecting for
detoxification mechanisms that promote resistance to two pesticides, which is referred to as cross resistance (as mentioned
previously).

The rotation of pesticides will only be effective in delay-
ing the development of resistance if the pesticides used select
for different resistance mechanisms (described previously),
although there is no way to control for this. For example, metabolic mechanisms might confer resistance to pesticides in different chemical classes that have different modes of action. As
such, rotation programs should encompass as many pesticides
with discrete modes of action as possible. A thorough understanding of resistance, and the factors—both operational and
biological—that can enhance resistance developing in insect or
mite pest populations is important. When rotating pesticides
with different modes of action, the assumption is that the frequency of individuals resistant to one pesticide will diminish
during the application of other pesticides. Nonetheless, issues
associated with resistance can be alleviated by implementing
nonpesticidal practices and rotating pesticides with different
modes of action, which is important because the exorbitant
costs associated with development and the regulations affiliated with registering new pesticides limits the number of new

pesticide active ingredients introduced into the marketplace. Therefore, always judiciously use pesticides currently available to avoid problems with pesticide resistance. Another consideration is that because consumers demand blemish-free plants or plants with no insect or mite pests, or visible damage, greenhouse producers are forced to make frequent pesticide applications, thus promoting resistance. If consumers would accept plants with insects and mites, and some damage, there would be less input from pesticides.

Multiple Pest Complexes

Greenhouse producers, in general, and depending on the types of horticultural crops being grown, usually contend with a multitude of insect or mite pests, not just one. Pesticide label rates can vary depending on the pest. However, using lower label rates or less than the recommended label rate may result in sublethal effects, which may increase the potential for resistance developing in pest populations. Another effect of using broad-spectrum pesticides or mixing pesticides together is the removal of competition, which may lead to pest problems. For example, western flower thrips will feed on populations of the twospotted spider mite. So disrupting western flower thrips populations with a pesticide application may lead to increased problems with the twospotted spider mite.

Many crops, including annual bedding plants and potted crops, such as chrysanthemum (*Dendranthema* x *grandiflorum*), are typically attacked simultaneously by several insect and/or mite pests including aphids, thrips, and whiteflies. So, a pesticide program developed for one pest must not directly affect the management strategies used to suppress populations of another pest. The management of one pest may interact with and influence the management of other insect or mite pests. There should be an emphasis on multiple pest complexes (including diseases) instead of simply concentrating on individual pest species. Another factor to consider in regards to multiple pest complexes is pesticide resistance. For instance, pesticide resistance may be exacerbated when western flower thrips populations are exposed to pesticide applications designed to target other pests. Botanical insecticides, for example, such as pyrethrins, used against other pests may lead to outbreaks of western flower thrips populations because resistant individuals survive and any natural enemies are eliminated. Also, many western flower thrips populations are known to be resistant to certain pyrethroid insecticides. Resistance may not only develop to pesticides targeting western flower thrips, but also to those used against other pests.

Broad-spectrum insecticides can disrupt management of western flower thrips or whiteflies, and also can impact the management of other pests including spider mites, whiteflies, and leafminers by eliminating natural enemies of these pests. Consequently, a problem with a secondary pest, such as spider mites and leafminers, can arise.

Another potential problem can occur when a pesticide has different label rates for two insect pests (e.g., western flower thrips and leafminers). As an example, the label rate may be higher for one insect pest than for another insect pest. So, if the higher label rate is used for one insect pest, the pesticide application may place selection pressure on the other insect pest population possibly increasing the rate of resistance. Therefore, always use pesticides that have similar label rates for two or more insect or mite pests.

Furthermore, fungicides and even plant growth regulators may profoundly impact pest management of insect and mite pests—either directly or indirectly. The isolation of individual insect or mite pests should be avoided and understanding the entire pest complex needs to be considered. Another means of dealing with multiple pest complexes is mixing together two pesticides to create pesticide mixtures.

Pesticide Mixtures

Pesticide mixtures or tank mixing typically involves combining two or more pesticides into a single spray solution. The mixture exposes individuals in a pest population to each pesticide simultaneously, with the intention of improving pest suppression. Although there are benefits associated with pesticide mixtures, several issues should be considered in advance. First of all, always read the pesticide label and understand why certain pesticides are being mixed together. Pesticide mixtures should be used that are appropriate based on the mode of action of each pesticide and development stage(s) of the target pest(s) for which the pesticide mixture is most effective. For example, tank mixing two pesticides (miticides) that only have activity on the adult stage of the twospotted spider mite would not be appropriate because both miticides would only kill adults and not larvae, nymphs, and/or eggs. However, tank mixing a pesticide with adult activity with another pesticide that is active on eggs, larvae, and nymphs would be appropriate because the pesticide mixture targets all life stages of the twospotted spider mite. What is most important is to always tank mix pesticides with different modes of action (refer to the section "Understanding Issues Associated with Pesticide Resistance").

One reason for mixing pesticides together is convenience because it is less time consuming, costly, and labor intensive to mix two or more pesticides, and make one application than to make separate applications of each pesticide. Another benefit of pesticide mixtures is the potential for improved pest suppression. For example, tank mixing two pesticides may result in greater mortality than applying either pesticide separately, which is referred to as synergism or potentiation. *Synergism* is when the combined toxicity of two pesticides is greater than the sum of the toxicities of each individual pesticide. Another case is when one compound has pesticidal activity but the other compound in the mixture has minimal if any toxicity when applied separately. The latter compound is usually a synergist (described below). *Potentiation* occurs when the activity of one pesticide enhances the activity of another pesticide. In this instance, both compounds have pesticidal activity, which may be toxic when applied separately.

Some compounds are synergists, which are adjuvants that enhance the effectiveness of the pesticide active ingredient. For example, piperonyl butoxide (PBO), which is not a pesticide, is commonly mixed or formulated with pyrethrins [botanical pesticides derived from chrysanthemum (*Chrysanthemum cinerariaefolium*) flowers] and certain pyrethroid pesticides. PBO works by blocking enzymes in the insect that detoxify the active ingredient so the pesticide no longer has activity. Furthermore, certain organophosphate pesticides are useful synergists for pyrethroid pesticides because they bind to particular enzymes responsible for detoxification, thus counteracting the ability of insect pests to develop resistance, which may be why pesticide manufacturers formulate organophosphate and pyrethroid pesticide mixtures to manage insect and mite pests.

There are issues associated with pesticide mixtures. For instance, just as synergism may improve the efficacy of pesticide mixtures, the opposite may occur when mixing two or more pesticides reduces the effectiveness of a pesticide mixture compared to separate applications of each pesticide, which is referred to as *antagonism*. In addition, applying pesticides together may not only reduce effectiveness, but may cause plant injury (such as phytotoxicity). Therefore, always read the labels of all pesticides to be mixed together in order to determine if pesticides should be mixed together. Contact the pesticide manufacturer directly if any questions arise.

Another potential issue with pesticide mixtures is incompatibility, which is a physical condition that prevents pesticides from mixing properly in a spray solution, thus reducing effectiveness or increasing the potential for plant injury (such as phytotoxicity). Incompatibility may be due to the chemical or physical nature of the pesticide(s), water impurities, water temperature, or formulation types mixed together. A way to determine incompatibility between two (or more) pesticides is to conduct a jar test, which involves collecting a sample of the spray solution into a separate empty container, and allowing the solution to remain idle for approximately 15 minutes (Figure 6.8). If the pesticides are not compatible, there will be a noticeable separation or layering, or flakes or crystals will form. However, if the pesticides are compatible, then the solution will appear homogeneous or resemble milk. Regardless, this procedure only assesses compatibility—not synergism or antagonism.

New plant varieties or cultivars are continually being introduced into the marketplace. However, these new varieties or cultivars may differ in their response to pesticide mixtures. Therefore, in order to avoid problems associated with plant injury (such as phytotoxicity), always test a pesticide mixture by making an application to about 10 plants before exposing the entire crop to the pesticide mixture. If no plant injury

Figure 6.8 Jar test to determine pesticide compatibility.

is apparent, then the pesticide mixture can be applied to the main crop.

The issue regarding pesticide mixtures and resistance is still not well understood, although applying two or more pesticides at different intervals may offer similar advantages as a pesticide mixture. However, using pesticide mixtures to mitigate resistance development may not be appropriate because each individual pest in the population may not receive a lethal dose or concentration of each pesticide. Nonetheless, the mixing of pesticides with different modes of action might delay resistance within a given pest population because mechanisms needed to resist the pesticide mixture may not be widespread or even exist in the pest population. Furthermore, there may be difficulties associated with individuals in the pest population to develop resistance to several modes of action simultaneously. In addition, individuals in the pest population resistant to one or more pesticides would likely succumb to the other pesticide in the mixture.

The ability of insect and mite pest populations to evolve resistance depends on a number of factors, including previous exposure to pesticides with similar modes of action. Moreover, using pesticide mixtures to mitigate resistance will only be effective if there is no cross resistance (see the section "Understanding Issues Associated with Pesticide Resistance") among individuals in the pest population to any of the pesticides in the mixture.

Pesticide mixtures have both advantages and limitations. Although greenhouse producers mix pesticides (this includes insecticides, miticides, and fungicides) together in order to reduce labor costs affiliated with multiple spray applications and to improve suppression of insect and mite pests, always exercise caution or avoid using pesticide mixtures so as to prevent problems associated with antagonism, incompatibility, phytotoxicity, and resistance.

Chapter 7

Biological Control

A biological approach involves using natural enemies (or biological control agents) such as parasitoids (parasitic wasps), predators, and/or pathogens (entomopathogenic nematodes and fungi) to suppress or regulate insect and mite pest populations in greenhouses. It is important to understand that biological control is a regulatory process; natural enemies will not eradicate an insect or mite pest population. The success of natural enemies is based primarily on their ability to maintain insect or mite pest numbers at levels low enough to minimize plant damage and subsequently maintain their own populations. The reason why greenhouse producers are interested in implementing a biological control program is due to issues associated with pesticide resistance, in which the efficacy of a given pesticide that targets a particular insect and/or mite pest population is substantially less than what was obtained previously. However, there is a general perception that biological control costs more than using pesticides, although this is not necessarily the case. When factors such a resistance, disposal, worker safety, and plant safety associated with pesticides are taken into consideration then biological control may actually be less expensive than pesticides.

The advantages of implementing a biological control program in greenhouses are: (1) fewer worker and customer exposure risks associated with pesticide residues on plants, (2) fewer rules and regulations (compared to pesticides), (3) no plant injury (such as phytotoxicity) or safety issues, (4) no issues affiliated with pesticide resistance, (5) minimal equipment required for application, and (6) marginal cleanup needed after application. However, there are potential disadvantages associated with a biological control program including: (1) inconsistent regulation of insect or mite pest populations; (2) inconsistent availability of natural enemies; (3) specificity, which is a problem when dealing with multiple insect and/or mite pests simultaneously; (4) cost of natural enemies, which is usually related to shipping; (5) short shelf life; and (6) quality control issues related to natural enemies (poor quality natural enemies will not be able to provide sufficient regulation of pest populations).

Quality Control

Since quality control is such a critical component in the success of a biological control program it is important to elaborate on this topic. In fact, the success of any biological control program is contingent on the quality of natural enemies released. The quality of natural enemies is dependent on rearing conditions, packaging, survival in transit, and handling by the end user (greenhouse producer). Natural enemies are normally reared in large quantities as colonies (population of individuals) under laboratory conditions where they feed on a food source such as a sugar-water solution (honey water) or prey that is feeding on plants.

However, there are a number of problems that can be encountered when rearing natural enemies. First, molds or diseases may inadvertently contaminate colonies, thus reducing viability and colony vigor. Second, the vigor and fitness of natural enemies including parasitoids may be affected during artificial rearing conditions because the environment (temperature, photoperiod, and relative humidity) under laboratory conditions can select for individuals that are more likely to succeed. However, when natural enemies are released, the environmental conditions in the greenhouse may be very different thus leading to shorter life spans. Third, parasitoids may not be able to locate prey because parasitoids were reared in cages where finding prey is easy, thus negatively affecting their foraging behavior. Fourth, continual rearing of parasitoids may favor sex ratios that are skewed toward females or males, and parasitoids reared and adapted to laboratory conditions may mate more frequently with females being more fecund (having more eggs). Nonetheless, the conditions in greenhouses may be entirely different so that these circumstances do not exist, which can negatively impact the long-term success of a biological control program because natural enemies cannot maintain pest populations below damaging levels. Fifth, during the process of rearing natural enemies, hyperparasites may be inadvertently introduced, when wild types are interposed into a colony. Hyperparasites are parasitoids that use parasitoids as a host, and the presence of hyperparasites can reduce the quality of a natural enemy colony.

Natural enemies are typically collected from established rearing colonies and then shipped to the end user (greenhouse producer). Biological control companies, in general, are responsible for producing, supplying, and distributing natural enemies. These can be producers/suppliers, distributors, or both. Distributors, in general, do not rear their own natural enemies; they order natural enemies from a producer/supplier. Producers have the facilities to rear one or more different types of natural enemies, which are subsequently sent to distributors upon request. Rearing natural enemies is expensive due to the space and labor involved, which is especially the case when natural enemies are being reared on prey feeding on plants or an alternative food source. In fact, most natural enemies are reared on alternative food sources or prey feeding on plants. However, a less costly method of rearing natural enemies involves using artificial diets as a primary or supplemental food source. Although artificial diets may be cheaper (requiring less space), quality problems can occur due to dietary affects that consequently negatively affect the performance of natural enemies when released into a greenhouse, which may be due to modifications in the foraging behavior or an inability to detect chemical cues such as semiochemicals (chemicals associated with communication among insects). Long-term rearing under laboratory conditions, particularly when using artificial diets to rear hosts for parasitoids, may have detrimental effects that compromise their ability to provide sufficient regulation of targeted pests. For example, parasitoid fecundity is influenced by the quality of the food source. Adult parasitoids must be provided with adequate food that contains carbohydrates as a source of energy, which is especially important for flight. Furthermore, female parasitoids should have been mated and had a preoviposition period prior to release.

Differences between laboratory and greenhouse environments may result in substantial natural enemy variability. Although rearing natural enemies under laboratory conditions may be ideal in regards to space and ability to control environmental conditions (e.g., temperature, photoperiod, and relative humidity), problems related to natural enemy performance in greenhouses may occur if rearing conditions differ widely from those in which natural enemies are released. For instance, searching efficiency, mate selection, and dispersal may be negatively influenced by laboratory conditions. Also, natural enemies reared on unnatural prey may be affected in regards to both quality and effectiveness because of being supplied with inadequate nutrients or inability to recognize prey due to the absence of odors. Moreover, quality control of natural enemies is important during mass rearing in order to maintain genetic variability of the colony.

Diseases including fungi, bacteria, viruses, and protozoa may compromise the quality of mass-produced natural enemies—both parasitoids and predators. Diseases may be present and then expressed during stress or remain undetected in a colony for an extended period. Natural enemies that are infected with a disease typically have higher mortalities and/or reduced performance compared to natural enemies that are disease-free. Releasing natural enemies that are infected with a pathogen may result in inadequate regulation. For example, the predatory mite, *Phytoseiulus persimilis*, when infected with a protozoa, produces fewer eggs and has reduced survival rates. In addition, protozoa such as microsporidia (spore producing protozoa) can reduce the fecundity and foraging behavior of the predatory mite *Neoseiulus cucumeris*. The mass rearing of the whitefly parasitoid *Encarisa formosa* may increase susceptibility to microsporidia thus decreasing the ability of the parasitoid to regulate whitefly populations. Continuous rearing under laboratory conditions may create temporary starvation or localized overcrowding, which places stress on individual natural enemies thus increasing susceptibility to diseases. Crowding of parasitoids may create stressful conditions that amplify susceptibility to fungi and bacteria. Natural enemies infected with diseases may experience a decrease in fecundity, longevity, foraging behavior, and/or lower attack rates. In addition, natural enemies under stress due to inadequate rearing conditions may not reproduce and are subsequently more susceptible to diseases.

Even if natural enemies leave the producer/supplier (or insectary) in good condition, inappropriate shipping and/or handling procedures by the distributors or end users (greenhouse producers) may result in decreased quality of natural enemies before release. For example, inappropriate handling or exposure to extremes of hot or cold can negatively impact the performance of natural enemies when released into the greenhouse. Poor shipping conditions can result in receiving dead or injured natural enemies, which impacts the success of a biological control program. Shipping containers must provide adequate environmental conditions that are conducive for survival including appropriate temperature and relative humidity, but still allow for sufficient air exchange. Natural enemies should be delivered in a container that is packed with either Styrofoam peanuts or newspaper to minimize movement during transit (Figure 7.1). An ice pack should also be placed in the container to keep the natural enemies cool during the shipping process. Natural enemies that are improperly packaged may experience stress thus reducing their effectiveness. Also, natural enemies may not be packaged properly resulting in the end user (greenhouse producer) receiving dead or low numbers of live natural enemies.

During shipment, many natural enemies do not have an abundant food supply. In fact, some natural enemies may be shipped without food, which can consequently decrease survival rates unless natural enemies are shipped as eggs or pupae. Natural enemies must be shipped by a reliable carrier so that they are received the next day. An extended shipment time may result in higher mortality of natural enemies or a reduction in fitness, subsequently reducing their ability

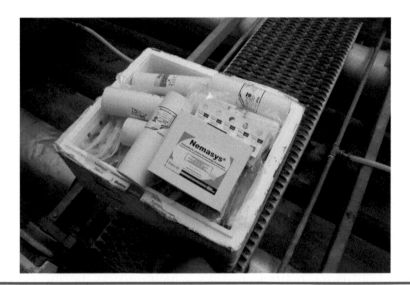

Figure 7.1 Container of natural enemies (biological control agents).

to provide sufficient regulation of insect or mite pest populations. Any delays during shipment may lead to mortality due to cannibalism (eating each other) or desiccation (drying up). For instance, when packaged or shipped in granular carriers, populations of the predatory mite *Phytoseiulus persimilis* are susceptible to cannibalism or desiccation when delays occur during shipping. Furthermore, *Phytoseiulus persimilis* adults that are shipped without prey and experience delays during shipment will start feeding on eggs and nymphs.

Parasitoids and predators should be stored for only a short period upon receipt. Storage of natural enemies for too long (e.g., over 7 days) can substantially decrease their fitness and ability to forage. Natural enemies stored in the adult stage may be more susceptible to reductions in fitness than those stored as immatures or pupae. Natural enemies, in general, should be released immediately upon arrival. Moreover, after natural enemies have arrived, always check them prior to release to insure the natural enemies are alive. For example, make sure adult parasitoids are flying around or predators shipped as adults or larvae are actively moving. If poor quality natural enemies are released into the greenhouse, inadequate regulation of insect or mite pest populations will likely result and consequently lead to frustration by the end user (greenhouse producer), which will be incorrectly perceived as the natural enemies not working.

In order to determine the quality of predatory mites such as *Phytoseiulus persimilis* that are typically shipped in containers consisting of bran or vermiculite, place a small sample on a white sheet of paper [21.6 cm × 27.9 cm (8.0 in. × 11.0 in.)] and check, with a 10× hand lens, to assess that the predatory mites are active. Natural enemies shipped as eggs or pupae may be evaluated using a different method. For instance, the quality of whitefly parasitoids such as *Encarsia formosa* and *Eretmocerus eremicus*, which are shipped on paper cards containing parasitized whitefly pupae (Figures 7.2 and 7.3), can be assessed by placing a sample card inside a glass or plastic jar with a lid, and regularly monitoring to insure that adults are emerging

Figure 7.2   Card with parasitized whitefly pupae that is attached to plant.

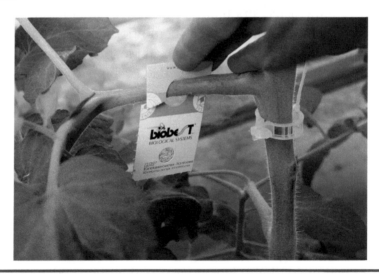

Figure 7.3   Placement of card with whitefly pupae parasitized by *Eretmocerus eremicus*.

(Figure 7.4). The actual number of parasitoids that emerged can be determined later when they all die in the container. Always assess the viability of entomopathogenic or beneficial nematodes (*Steinernema feltiae*) prior to use, which may be accomplished by taking a glass jar or beaker and removing a

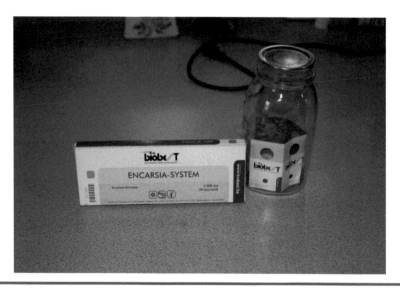

Figure 7.4   **Example of evaluating quality of natural enemy (in this case, the greenhouse whitefly parasitoid, *Encarsia formosa*).**

small sample from the nematode solution. Then hold the jar or beaker up to a light source or sunlight and look for nematodes actively moving around in the solution. If the nematodes appear straight or rigid and are simply floating around in the solution with no evidence of movement, then they are dead.

How to Implement a Biological Control Program

When making the decision to implement a biological control program it is important to: (1) determine the availability of natural enemies from suppliers and distributors; (2) understand the life cycle, biology, and behavior of insect and/or mite pests that feed on greenhouse-grown horticultural crops; (3) scout crops regularly (refer to Chapter 3, "Scouting") and order natural enemies early (at least 2 weeks in advance); and (4) conduct yearly follow-up assessments to determine successes and failures. The first procedure that needs to be considered (and implemented) before attempting biological control is the establishment of a reliable scouting program (refer to Chapter 3, "Scouting"). Moreover, in order to succeed with biological control always make sure to: (1) correctly identify all primary insect and mite pests (e.g., aphids, mites, thrips, and whiteflies); (2) identify available natural enemies for specific insect and mite pests; (3) establish a good relationship with a reliable supplier/distributor of natural enemies; (4) minimize pesticide residues; (5) start with a clean greenhouse by removing weeds and plant debris from both inside and outside the greenhouse (refer to Chapter 4, "Cultural Control and Sanitation"); (6) order natural enemies early (at least 2 weeks prior to release); (7) assess the quality of natural enemies by making sure they are alive upon receipt (see the section "Quality Control"); (8) release natural enemies immediately upon arrival; (9) apply natural enemies in the early morning or evening; (10) introduce or release natural enemies before insect and/or mite pest populations are abundant or reach outbreak proportions; (11) apply natural enemies at recommended release rates; (12) release enough natural enemies when the most susceptible life stage(s) are present, which will result in successful regulation of insect and/or mite pest populations; and (13) evaluate performance of natural enemies throughout the growing season. Reasons why biological control programs fail are as follows:

1. Improper identification of pest (insect or mite)
2. Not scouting
3. Not ordering the appropriate natural enemy (parasitoid or predator)
4. Not checking the quality of natural enemies upon receipt
5. Not releasing natural enemies immediately upon receipt
6. Making releases of natural enemies too late
7. Not releasing enough natural enemies
8. Applying too many natural enemies
9. Releasing natural enemies after applying a long-residual pesticide

Biological Control Approaches

The common type of biological control implemented in greenhouses is augmentation, which is a practice designed to increase the number of natural enemies by purchasing them from a supplier/distributor (Figures 7.5 and 7.6) and releasing them into the greenhouse. There are two approaches: *inundation* and *inoculation*. *Inundation* refers to releasing many natural enemies (of the same species) in order to quickly suppress an insect or mite pest population without relying on the subsequent progeny or next generation to provide any level of suppression. *Inoculation* is the release of small numbers of natural enemies conducted over a prolonged period with the progeny or next generation expected to provide some level of regulation or suppression.

Types of Natural Enemies

There are two distinct categories of natural enemies: specialist and generalist. Specialist natural enemies, which are usually parasitoids (Figure 7.7), feed on or attack only one insect or

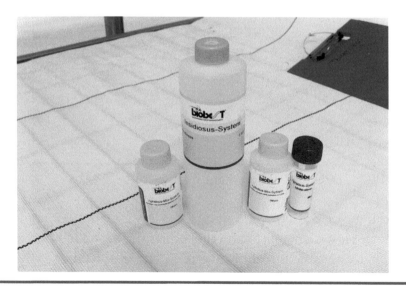

Figure 7.5 **Example of different commercially available natural enemies.**

Figure 7.6   Containers of commercially available natural enemies from suppliers/distributors.

Figure 7.7   Adult aphid parasitoid.

mite prey, or particular life stage (egg, larva, nymph, or adult) of a given prey. Generalist natural enemies, however, feed on or attack a variety of insect or mite prey, and also feed on different life stages (egg, larva, nymph, and/or adult) of a particular prey. Natural enemies, in general, must have good searching efficiency and be able to locate prey in patches or localized areas on plants. Furthermore, in order for natural enemies to succeed, they must be able to reproduce and have similar developmental times (from egg to adult) as the subsequent prey.

A parasitoid adult female can insert an egg into or on the body of an insect pest, which hatches into a young larva that consumes the internal contents. Parasitoids that attack aphids cause the aphids to expand in size and turn a light brown color. These are subsequently referred to as "mummified" aphids (Figures 7.8 and 7.9). Eventually, the parasitoid larva pupates and develops into an adult that creates an opening in the dead insect (Figure 7.10). The adult emerges, mates, and then the female disperses to attack other insect pests within the surrounding area (Figures 7.11 and 7.12). Parasitoids do not kill insect pests immediately. In general, most parasitoids are specific in regards to insect pest species and may only prefer certain life stages. The characteristics of parasitoids include: (1) kill the prey they are living off of; (2) free living in the adult stage with the immature either inside (endoparasitoid) or outside (ectoparasitoid) of prey; (3) kill prey slowly,

Figure 7.8 **"Mummified" or parasitized aphids on the underside of leaf.**

Figure 7.9   "Mummified" or parasitized aphids on leaf underside.

Figure 7.10   Close-up of "mummified" or parasitized aphids (note hole in back of dead aphid where the adult parasitoid emerged).

but may reduce prey fitness, feeding, and/or reproduction; (4) kill one prey with a single prey required for development although an individual female may be capable of laying over 100 eggs; and (5) specific in regards to insect species and particular life stage attacked. Parasitoids can be purchased from

Figure 7.11 **Parasitoid adult (top) and "mummified" aphid (bottom).**

Figure 7.12 **Parasitoid on leaf.**

commercial suppliers/distributors that are shipped in containers (Figure 7.13), which are placed among the crop in the greenhouse or on cards that are attached to plants (Figure 7.14).

Predators consume portions of or the entire insect or mite pest. Furthermore, many predators (Figures 7.15 and 7.16),

Figure 7.13   Container of the aphid parasitoid, *Aphidius colemani*. This parasitoid attacks both green peach and melon aphid.

Figure 7.14   Parasitized whitefly pupae on card that is attached to plant.

in general, feed on all life stages including eggs, larvae or nymphs, pupae, and adults. Characteristics of predators are: (1) feed on and thus kill more than one prey; (2) most life stages are free living (no pupal stage); (3) kill or consume prey quickly; (4) in general, both the immature and adult are

Figure 7.15   Close-up of minute pirate bug (*Orius* spp.) adult.

Figure 7.16   Close-up of ladybird beetle larva.

predacious; and (5) feed on a diversity of prey types depending on the specific predator.

In regards to predatory mites, there are three types: I, II, and III. Type I are specialist predatory mites that only feed and survive on spider mites in the Tetranychidae family

(e.g., twospotted spider mite). An example of a Type I predatory mite is *Phytoseiulus persimilis*. Type II includes selective predatory mites, such as *Neoseiulus californicus* and
Neoseiulus cucumeris, that feed on a broad range of prey
and will also feed on pollen. Type III are predatory mites
that feed on a very broad range of prey including eriophyid
mites (cigar-shaped mites with only four legs), and broad and
cyclamen mites. Examples of Type III predatory mites include
Amblyseius swirskii and *Neoseiulus californicus* (some predatory mites may overlap in regards to type). There are predatory
mites for use against aboveground pests (western flower thrips
and the twospotted spider mite) and belowground pests (fungus gnats). Predatory mites may be purchased from commercial suppliers/distributors in containers (Figures 7.17 and 7.18),
or in sachets (Figure 7.19) or holding packets (Figure 7.20)
that are attached directly to plants or containers (Figure 7.21),
which may be useful for hanging baskets.

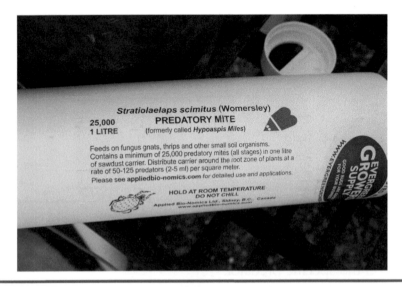

**Figure 7.17   Container of the predatory mite, *Stratiolaelaps scimitus*
(formally *Hypoaspis miles*). This predatory mite feeds on fungus gnat
larvae.**

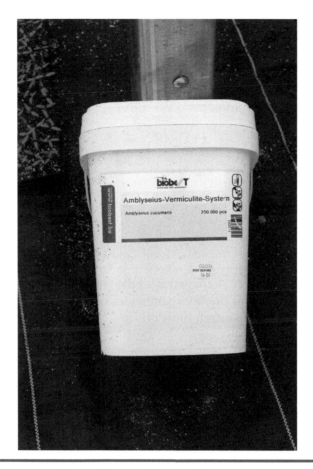

Figure 7.18 **Container of the predatory mite, *Neoseiulus cucumeris*.** This predatory mite feeds on the first larval instar of western flower thrips.

Entomopathogenic or beneficial nematodes are microscopic roundworms that enter the larval stage of insects through natural openings, such as the mouth, anus, or breathing pores (spiracles). Once inside the larva, the nematode releases a bacterium that attacks the midgut by producing protein destroying enzymes resulting in septicemia within 24–48 hours. Beneficial nematodes require moisture for survival, so they are mainly used against fungus gnat larvae residing in the growing medium. A commonly used entomopathogenic nematode is *Steinernema feltiae* (Figure 7.22). A list of commercially available

Figure 7.19 Sachet containing the predatory mite *Amblyseius swirskii* attached to plant. This predatory mite feeds on both the first and second larval instars of western flower thrips.

Figure 7.20 Holding packet of predatory mites.

natural enemies for use in greenhouse production systems against the major insect and mite pests is presented in Table 7.1.

Entomopathogenic or beneficial fungi work in a similar manner as parasitoids using the insect pest as a food source by consuming the internal contents. Entomopathogenic fungi

Figure 7.21   Sachets of predatory mite attached to containers of hanging baskets.

Nemasys®

Advanced Biocontrol for Horticulture

For control of Fungus Gnats (*Bradysia spp.*), and Western Flower Thrips (*Frankliniella occidentalis*) in greenhouse horticulture production. Contains infective juveniles of *Steinernema feltiae* nematodes in an inert water-soluble carrier.

Net contents: 50 million nematodes

KEEP OUT OF REACH OF CHILDREN
CAUTION

Figure 7.22   Entomopathogenic nematode product containing *Steinernema feltiae*.

are typically applied as sprays to the plant foliage. They penetrate through the insect skin (cuticle) by means of enzyme degradation or mechanical pressure. Once inside the insect, the entomopathogenic fungus colonizes the body cavity where the blood flows (hemocoel), which compromises the integrity of

Table 7.1   List of Biological Control Agents for Use in Greenhouse Production Systems against the Major Insect and Mite Pests

Pest	Biological Control Agents	Type	Comments
Aphids	*Hippodamia convergens*	Predator	Repeated releases are usually needed. Requires high aphid numbers to be effective.
	Chrysoperla rufilabris	Predator	A single larva may consume up to 300 aphids.
	Chrysopa carnea	Predator	May also feed on thrips, whiteflies, and mealybugs.
	Aphidoletes aphidomyza	Predator	May enter a diapause stage under short daylengths.
	Adalia bipunctata	Predator	Both larvae and adult feed on many different aphid species.
	Aphidius colemani	Parasitoid	Attacks both green peach and melon aphid.
	Aphidius ervi	Parasitoid	Attacks both foxglove and potato aphid.
	Aphidius matricariae	Parasitoid	Attacks the green peach aphid.
	Aphelinus abdominalis	Parasitoid	Attacks many different aphid species and adults will host feed.

(Continued)

Table 7.1 (*Continued*)   **List of Biological Control Agents for Use in Greenhouse Production Systems against the Major Insect and Mite Pests**

Pest	Biological Control Agents	Type	Comments
Fungus Gnats	*Stratiolaelaps scimitus*	Predator	Feeds on eggs, larvae, and pupae. Survives up to 7 weeks without a food source.
	Steinernema feltiae	Beneficial Nematode	Must be used early in the production cycle. Can be applied through chemigation systems.
	Dalotia coriaria	Predator	Feeds on both eggs and larvae. Adults will disperse throughout the greenhouse.
Leafminers	*Diglyphus isaea*	Parasitoid	Female lays eggs near larva and will also host feed. Requires high populations to be effective.
Mealybugs	*Cryptolaemus montrouzieri*	Predator	Less effective against the long-tailed mealybug.
	Leptomastix dactylopii	Parasitoid	Only attacks third to early fourth instar larvae of the citrus mealybug.
	Anagyrus pseudococci	Parasitoid	Attacks second instar and adult life stages of several mealybug species.

(*Continued*)

Table 7.1 (*Continued*)   List of Biological Control Agents for Use in Greenhouse Production Systems against the Major Insect and Mite Pests

Pest	Biological Control Agents	Type	Comments
Spider Mites	*Phytoseiulus persimilis*	Predator	Primarily attacks twospotted spider mite. Requires certain temperatures and relative humidity.
	Amblyseius fallacis	Predator	Tolerates low temperatures and feeds on pollen in the absence of prey.
	Amblyseius californicus	Predator	Tolerates high temperatures and low relative humidity. Feeds on pollen in the absence of prey.
	Amblyseius andersonii	Predator	Feeds on a wide range of spider mites and other prey as a food source.
	Galendromus occidentalis	Predator	Tolerates high temperatures and low relative humidity.
	Feltiella acarisuga	Predator	Adults do not feed. Females deposit eggs near areas infested with spider mites.

(Continued)

Table 7.1 (*Continued*)   List of Biological Control Agents for Use in Greenhouse Production Systems against the Major Insect and Mite Pests

Pest	Biological Control Agents	Type	Comments
Spider Mites	*Stethorus punctillum*	Predator	Feeds on all life stages of spider mites. Adults will disperse to locate spider mite infestations.
Shore Flies	*Dalotia coriaria*	Predator	Adults and larvae are predaceous. Adults are very mobile and will disperse within a greenhouse.
Thrips	*Neoseiulus cucumeris*	Predator	Only attacks first instar larva. Releases should be made early in the production cycle.
	Amblyseius swirskii	Predator	Attacks both first and second instar larvae.
	Stratiolaelaps scimitus	Predator	May attack pupae located in the growing medium.
	Steinernema feltiae	Beneficial Nematode	May be effective on pupae in the growing medium.
	Orius insidiosus	Predator	Feeds on both larvae and adults.
	Dalotia coriaria	Predator	May feed on pupae in the growing medium.

(Continued)

Table 7.1 (*Continued*)   List of Biological Control Agents for Use in Greenhouse Production Systems against the Major Insect and Mite Pests

Pest	Biological Control Agents	Type	Comments
Whiteflies	*Encarsia formosa*	Parasitoid	Primarily used against the greenhouse whitefly. Host feeding by adult may kill whitefly larvae.
	Eretmocerus eremicus	Parasitoid	Primarily used against the sweet potato whitefly.
	Amblyseius swirskii	Predator	Feeds on both eggs and nymphs. Will also feed on pollen as an alternative food source.
	Delphastus catalinae	Predator	Feeds on both eggs and nymphs. Useful for dealing with high whitefly populations.

the insect immune system. Death may occur after 3–14 days of exposure. Insect death is usually dependent on the dose, with higher fungal spore concentrations resulting in faster kill and subsequently higher mortality. Examples of entomopathogenic fungi that can be used in greenhouses include *Beauveria bassiana*, *Metarhizium anisopliae*, and *Isaria fumosoroseus*.

Banker Plant System

Banker plants are used to assist in the development and dispersal of natural enemies including parasitoids and predators for regulation of plant-feeding insects and mites.

The primary purpose of banker plants is to rear prey and always have a continually reproducing and prolonged population of natural enemies. There are two types of banker plant systems. One involves using the same insect (or mite) pest that is already present in the greenhouse. The other one uses an alternate prey feeding on a plant that is not grown in the greenhouse such as winter barley (*Hordeum vulgare*) or winter wheat (*Triticum* spp.). The alternate prey is what the natural enemy feeds on and uses for reproduction. A commonly used alternate prey is cereal aphids, including the corn-leaf aphid (*Rhopalosiphum maidis*) and bird-cherry aphid (*Rhopalosiphum padi*).

The success of using banker plants is contingent on location. Banker plants need to be placed along walkways and at the ends of benches or walkways (Figures 7.23 and 7.24). For example, studies evaluating the dispersal of aphid parasitoids have reported that *Aphidius colemani* tends to migrate 0.97–1.98 m (3.2–6.5 ft.) from the point of release and that the highest parasitism rates occur within only a few meters from the point of release. Therefore, evenly distribute banker plants throughout the greenhouse. General recommendations suggest placing banker plants approximately 40 m (131 ft.) apart in order to enhance parasitization or to use 4–5 banker plants per 929 m² (10,000 ft.²).

Common rye (*Secale cereale*) plants infested with aphids are typically used in banker plant systems. For instance, rye plants with the bird-cherry aphid can serve as banker plants for the aphid parasitoid *Aphidius colemani*. This parasitoid and others may attack aphid species including the green peach, melon, foxglove, and potato aphid. The disadvantage of using rye as a banker plant is that rye is susceptible to western flower thrips and therefore may serve as a reservoir for populations of the western flower thrips. Barley and wheat are used in banker plant systems designed to regulate aphid populations by providing refuge and harboring prey for *A. colemani* and the aphid predator *Aphidoletes aphidimyza*.

Figure 7.23   Banker plant placed on greenhouse bench. In this case, the banker plant is used to retain parasitoids of pest aphids that are located on the main crop.

Figure 7.24   Banker plants placed on floor near walkway. Banker plants are used to retain parasitoids of pests aphids, which are located on the main crop.

Another banker plant system involves allowing the predatory bug *Dicyphus hesperus* to develop on common mullein (*Verbascum thapsus*) plants that are placed among tomato (*Solanum lycopersicum*) plants grown in greenhouses for regulation of greenhouse whitefly populations. Banker plants can also be used to rear predators of the twospotted spider mite. For example, bush or snap bean (*Phaseolus vulgaris*) plants infested with twospotted spider mites serve as banker plants for predators such as the predatory mite *Phytoseiulus persimilis* and the predatory midge *Feltiella acarisuga*. Plants are distributed in the greenhouse among crops where twospotted spider mite populations are commonly located. However, the use of bush or snap beans may lead to potential problems with western flower thrips populations that migrate from banker plants to tomatoes, and extensive pest populations may develop on bush or snap bean plants that are neglected.

Banker plant systems have been successful in situations where corn (*Zea mays*) plants contain populations of the Banks grass mite (*Oligonychus pratensis*), which serve as prey for the predatory mite *Neoseiulus californicus*. The corn plants need to be distributed among the main crop(s) so that the predatory mites can migrate back and forth from the banker plants to the main crop(s) and vice versa. In addition, varieties of ornamental pepper (*Capsicum annuum*), when used as banker plants, support populations of the predatory mite *Amblyseius swirskii*, which subsequently provide adequate suppression of populations of the silverleaf whitefly (*Bemisia tabaci* biotype B), western flower thrips, and chilli thrips (*Scirtothrips dorsalis*) under greenhouse conditions.

The basic components of a banker plant system include: (1) the banker plant, (2) the natural enemy, and (3) the alternate prey. A benefit of using banker plants in greenhouses is that banker plants maintain populations of natural enemies over an extended period, which may reduce the need to purchase additional natural enemies. However, banker plants should be used in conjunction with other pest management strategies. Always consult with a biological control supplier/distributor for more information on how to successfully utilize banker plants.

Effect of Environmental Conditions on Natural Enemies

Environmental conditions in greenhouses including temperature, relative humidity, photoperiod, and light intensity can impact the effectiveness of natural enemies in regulating pest populations by influencing fecundity, foraging behavior, and survival. Therefore, the environment must be taken into consideration when implementing a biological control program. For instance, temperature influences the rate of development and reproduction of both natural enemies and pests. In addition, low light intensity may decrease the activity of certain natural enemies. Natural enemies have to compete with pest population development and reproduction in order to successfully regulate populations. In fact, the intrinsic rate of increase (rate of increase of populations that reproduce within discrete time intervals and have generations that fail to overlap) of natural enemies must be similar to pests in order to provide sufficient regulation.

The predatory mite *Phytoseiulus persimilis* is most effective at temperatures between 20°C and 30°C (68°F and 86°F). Searching activity decreases when temperatures are >30°C (86°F). However, at temperatures >30°C (86°F), the development time of the twospotted spider mite is shorter than *Phytoseiulus persimilis*. Furthermore, a relative humidity <40% may affect egg survival, adult longevity, and female fecundity of the predatory mite. Excessive light intensity causes *Phytoseiulus persimilis* to inhabit the underside of leaves and lower plant canopy, which may allow twospotted spider mite populations to escape attack by feeding on the upper portions of the plant.

Relative humidity may impact the ability of parasitoids to attack prey. For instance, the whitefly parasitoid *Encarsia formosa* has a higher parasitism rate at a relative humidity between 50% and 70%. Moreover, daylength and light intensity may influence natural enemies. The aphid predator *Aphidoletes aphidimyza*, for example, enters a resting (diapause) stage when experiencing short daylengths (<12 hours). The predatory mite *Neoseiulus cucumeris* enters a diapause stage during the winter when temperatures are <21°C (70°F) and the daylength is <15 hours. In addition, *Encarsia formosa* may be less effective during the fall months due to the direct effects of lower light intensities. Reports indicate that the parasitoid *Eretmocerus eremicus* is more effective than *Encarsia formosa* in regulating populations of the greenhouse whitefly under lower light intensities. Therefore, more frequent release rates of *Encarsia formosa* may be required in order to compensate for this condition; however, costs associated with purchasing the natural enemy and making frequent releases may be too expensive.

Effects of Plants on Natural Enemies

Plants produce two types of defenses: extrinsic and intrinsic. *Extrinsic defenses* are associated specifically with natural enemies in which plants produce and then emit volatile compounds (e.g., terpenoids) to attract natural enemies that will attack plant-feeding insects and mites. Plants may respond to insect and mite feeding by releasing volatiles from damaged areas. These volatiles subsequently help natural enemies locate specific prey on areas of the plant. For example, lima bean (*Phaseolus* spp.) plants emit volatiles from leaves that are damaged by the twospotted spider mite to attract predatory mites, which increases the ability of predatory mites to locate prey on damaged plants as opposed to searching randomly. These volatile compounds can also help certain parasitoids searching for small prey, which are difficult to locate because they may be well camouflaged or are feeding on leaf undersides. Volatile compounds emitted from plant leaves in response to insect or mite feeding damage may allow parasitoids and predators to determine any differences between infested and noninfested plants, which increases the efficiency of natural enemies in locating prey. The concentration of volatiles emitted depends on plant type, insect or mite pest, and environmental conditions (e.g., daylength and light intensity). Furthermore, certain cultural conditions such as water stress may influence the release of volatiles from plant leaves.

Intrinsic defenses are produced by plants either chemically via toxins that affect digestion, or physically by means of impediments such as leaf toughness, cuticle thickness, leaf waxiness, and trichomes (leaf hairs). Physical defenses can increase the development time of larvae or nymphs, which enhances exposure and thus susceptibility to natural enemies. For instance, tough leaves or thick cuticles may decrease feeding rates and consequently extend the development time of the immature life stages, which may then increase exposure to natural enemies. However, natural enemies including ladybird beetle adults may fall off of plants with leaves having waxy cuticles, thus allowing aphids to escape attack.

Trichomes or hairs on leaves (foliar pubescence) can negatively affect the performance of parasitoids and predators by hindering movement, influencing walking speed, and modifying walking patterns. These behavioral responses may be associated with length, alignment, and/or density of trichomes. Trichomes can be angled downward forming a physical barrier; hooked, thus entangling natural enemies; or glandular, which entraps natural enemies by means of adhesion or natural enemies may be killed directly by contacting toxic fluids. Trichomes may also impact foraging behavior such as searching ability or efficiency of parasitoids or predators, which decreases their ability to regulate pest populations. Trichomes may directly impede the ability of certain natural enemies to locate prey on plants. Furthermore, plant leaves with a dense layer of trichomes inhibit the ability of natural enemies to move around. Natural enemies may also turn around more frequently on leaves with a high trichome density, leading to searching on areas of the leaf that were already visited.

Trichome density may negatively affect encounter rate, thus compromising the success of natural enemies in regulating pest populations. For example, larvae of the aphid predator *Coleomegilla maculata* fall off cucumber (*Cucumis sativus*) leaves that contain many trichomes, which reduces the ability of the predator to find aphids on plants. Transvaal daisy (*Gerbera jamesonii*) leaves and those of certain vegetable plants with high trichome densities may decrease the walking speed and predation rate of the predatory mites *Amblyseius swirskii*, *Neoseiulus cucumeris*, and *Phytoseiulus persimilis*, thus negatively affecting the ability of these predators to regulate pest populations. Honeydew, which is a clear, sticky liquid excreted by phloem-feeding insect pests such as aphids, mealybugs, soft scales, and whiteflies may accumulate more on leaves with high trichome densities than on smooth leaves or those without trichomes. Consequently, the presence of honeydew may lead to parasitoids and even predators encountering honeydew droplets, resulting in natural enemies spending more time preening/grooming themselves as opposed to searching for prey.

Certain cucumber varieties contain trichomes that interfere with the searching efficiency and parasitism rates of the parasitoid *Encarsia formosa* against the greenhouse whitefly because the parasitoid has difficulty finding and thus parasitizing greenhouse whitefly larvae on cucumber varieties with dense trichomes. The trichomes interfere with searching efficiency by negatively affecting walking speed and pattern of movement. Therefore, reducing the number of trichomes on leaves may increase the ability of *Encarsia formosa* to effectively regulate populations of the greenhouse whitefly. In fact, the parasitoid may walk faster on leaves with fewer trichomes resulting in greater searching efficiency on cucumber varieties with no trichomes compared to varieties with trichomes. Thus, *Encarsia formosa* can parasitize more greenhouse whitefly larvae on cucumbers with fewer trichomes than cucumbers with higher trichome densities.

Plant breeding programs should concentrate on developing plant cultivars that allow natural enemies to be more effective. For instance, plants with fewer trichomes may result in an increase in the searching efficiency of certain parasitoids including *Encarsia formosa*. The leaves of transvaal daisy (*Gerbera jamesonii*) may contain trichomes that impede the effectiveness of both *Encarsia formosa* (for the greenhouse whitefly) and the predatory mite *Phytoseiulus persimilis* (for the twospotted spider mite). However, *Encarsia formosa* may have difficulty on smooth leaves (no trichomes) due to walking so fast they fail to notice greenhouse whitefly larvae (nymphs), thus resulting in reduced parasitism rates. In addition, the increased walking speed may decrease the ability of the parasitoid to regulate greenhouse whitefly populations possibly leading to plant damage.

Plants such as tomato (*Solanum lycopersicum*) possess glandular trichomes that entangle or entrap natural enemies in a sticky exudate, or the exudate may accumulate on their bodies, consequently impairing movement, which may cause natural enemies to spend more time grooming themselves instead of searching for prey. The whitefly parasitoid *Eretmocerus eremicus* may become entrapped within the exudate of glandular trichomes of certain plant types, thus inhibiting their ability to locate prey, or killing them outright. Glandular trichomes of the plant *Nicotiana glutinosa* may entangle the parasitoid *Encarsia formosa*. Potato (*Solanum tuberosum*) plants with an abundance of glandular trichomes are less susceptible to aphids, which can negatively affect natural enemies of aphids as the natural enemies become entrapped in the exudates produced by the glandular trichomes.

Another factor to consider in regards to the effect of plants on natural enemies is the direct impact of flowers on the ability of natural enemies to regulate pest populations. For instance, the structural complexity (petals and sepals) of certain flower types may influence the success of natural enemies in regulating pest populations by providing refuge for prey, which subsequently allows them to escape exposure from natural enemies.

How to Successfully Release Natural Enemies

Release natural enemies early in the crop production cycle in order to maximize effectiveness, especially prior to insect and mite pest populations reaching outbreak proportions. As previously mentioned, natural enemies must be released immediately upon arrival because they have a very short shelf life. Furthermore, always make sure that natural enemies are alive before release in order to insure their effectiveness.

A basic understanding on how natural enemies respond to pest numbers under various environmental conditions helps in determining the development and reproduction of insect and mite pests, and the natural enemy, which influences the ability of natural enemies to sufficiently regulate pest populations. Moreover, knowledge of the following is essential: (1) what life stages of a particular pest are attacked by a parasitoid so that releases can be synchronized to coincide with pest developmental life stages (eggs, larvae, nymphs, pupae, and adults) that are susceptible to the parasitoid; (2) the spatial distribution of prey within a greenhouse, which can influence the searching ability of certain natural enemies and potential release rates; and (3) the number of parasitoids or predators to release based on the numbers of a particular pest in the crop. For instance, releasing too many natural enemies may result in mutual interference (interactions of natural enemies competing for the same food source). Releases of natural enemies should be initiated early and frequently enough during the crop production cycle to maintain pest populations below damaging levels. Also, be sure to remove any yellow sticky cards prior to releasing parasitoid or predator adults because the adults may be attracted to and subsequently captured on yellow sticky cards. Yellow sticky cards can be replaced 1 week after releasing natural enemies.

Pesticides and Natural Enemies

The sole use of natural enemies may not always be sufficient to suppress insect or mite pest populations in greenhouses due to the reproductive capacity of particular insect and mite pests, such as aphids and the twospotted spider mite. Therefore, another pest management strategy that has been considered is using pesticides (insecticides and miticides) in conjunction with natural enemies in order to manage certain insect or mite pests, which has been associated with using so-called selective pesticides (described in Chapter 6, "Pesticides"). Selective pesticides include insect growth regulators, insecticidal soaps and horticultural oils, selective feeding blockers, and microbial agents such as entomopathogenic fungi and bacteria, and related microorganisms. In general, selective pesticides may be less harmful to natural enemies than conventional broad-spectrum pesticides because selective pesticides are active on a narrower range of target pests than conventional pesticides. However, there are selective pesticides that may be harmful to certain natural enemies. Although selective pesticides, in general, may not cause direct or immediate harm to natural enemies, there may be indirect or even sublethal effects affiliated with using selective pesticides including delayed development of the prey and natural enemy, inhibition of adult emergence, and/or reduced natural enemy survival.

The harmful effects of pesticides may be associated with: (1) direct contact, (2) host elimination, (3) residual activity, and (4) indirect effects. Direct effects are affiliated with directed sprays that cause mortality of natural enemies outright. Host elimination involves mortality of prey, which leads to either natural enemies dying or leaving the area because they are unable to locate any remaining prey. Residual activity is related to spray applications not causing direct mortality of natural enemies but any remaining residues having repellent activity that negatively impacts the ability of parasitoids or predators to find a food source. Indirect effects occur when the pesticide does not cause direct mortality of a natural enemy but may affect reproduction (e.g., sterilizing females), decrease the ability of females to lay viable eggs, reduce adult longevity, affect the sex ratio (e.g., more males produced than females), and negatively influence the foraging behavior such that parasitoids and predators cannot locate prey. Some of the biological parameters associated with natural enemies that may be indirectly affected by pesticides are:

1. Longevity	8. Reproduction
2. Fecundity and/or fertility	9. Development time (egg to adult)
3. Mobility	10. Prey searching efficiency and feeding behavior
4. Predation and/or parasitism	11. Sex ratio
5. Emergence rates	12. Prey consumption
6. Population growth/reduction	13. Prey acceptance
7. Orientation behavior	

The indirect effects may be more subtle than direct effects although indirect effects may: (1) inhibit the ability of natural enemies to establish populations, (2) influence the capacity of natural enemies to use prey, (3) impact parasitism (for parasitoids) or consumption (for predators) rates, (4) diminish female reproductive capacity, (5) reduce prey availability, (6) inhibit the ability of natural enemies to recognize prey, and (7) reduce foraging behavior thus impacting locating prey. Indirect effects may vary depending on if the pesticide is contact, stomach poison, translaminar, or systemic.

Natural enemy susceptibility to selective pesticides may be influenced by a number of factors including: (1) the natural enemy type (e.g., parasitoid or predator), (2) natural enemy species, (3) natural enemy life stage (e.g., egg, larva, pupa, and adult), (4) development stage of prey, (5) application rate, (6) application timing, and (7) mode of action of selective pesticide. Another factor to consider is that any harmful effects associated with selective pesticides may not be related with the active ingredient but due to the inert ingredients including adjuvants (e.g., surfactants).

Selective pesticides, in general, are more specific in regards to the insect and/or mite pests they are active against and are less harmful to natural enemies than typical conventional pesticides. However, direct spray applications (wet sprays) of pesticides such as insecticidal soaps (potassium salts of fatty acids) and horticultural oils (mineral based) may be harmful to certain natural enemies, particularly parasitoids. Nonetheless, insecticidal soaps and horticultural oils tend to be less directly harmful than conventional pesticides, especially after residues have dissipated.

The simultaneous use of pesticides and natural enemies in pest management programs may be highly variable due to the multiple interactions (e.g., pest–natural enemy–pesticide). Interactions are primarily based on the type of selective pesticide, type of natural enemy (e.g., parasitoid or predator), and stage of development (e.g., egg, larva, nymph, pupa, and adult). The use of pesticides with natural enemies is most likely to be successful in systems with long-term horticultural crops such as greenhouse-grown cut flowers or seasonal potted plant crops, such as poinsettia (*Euphorbia pulcherrima*).

Issues Regarding the Use of Biological Control

Biological control can be a challenge in greenhouse-grown horticultural crops due to the polyculture system in which a diversity of crop types, and cultivars and varieties, are produced simultaneously. Some horticultural crops are extremely susceptible to pests, with pest development and reproduction so rapid that natural enemies cannot keep up, thus reducing their effectiveness in regulating pest populations. In addition, tolerance for insect and mite pests is extremely low resulting in pest populations that are sparsely distributed, which may negatively influence the performance of natural enemies. Furthermore, the high reproductive capacity of certain pests such as aphids and the twospotted spider mite may inhibit the ability of natural enemies to effectively regulate pest populations below damaging levels. Therefore, pesticides may be needed or releases of natural enemies would have to be initiated very early and frequently enough in order to successfully regulate aphid and twospotted spider mite populations. Moreover, biological control may not be feasible when dealing with insect pests like the western flower thrips that are capable of vectoring plant diseases. Regarding production, some crops such as annual bedding plants are grown and shipped within a short period (<6 weeks), which may negate the associated costs and even practicality of using biological control.

In some cases, biological control may not be a practical option as the cost of implementing a biological control program could be too expensive to justify. There may also be issues associated with using biological control in multiple pest complexes due to the logistics involved in releasing several natural enemies simultaneously, as the environmental conditions (e.g., temperature and photoperiod) and the interactions among the different natural enemies may reduce their effectiveness. One major limiting factor is consistent availability or supply of adequate numbers of natural enemies when purchased from distributors. Another potential concern is that some pests may still be present on plants prior to shipping, which could impact consumer acceptance and thus salability. However, even pesticides will not kill all pests. Therefore, greenhouse producers need to promote and educate customers regarding the benefits of using biological control instead of solely applying pesticides.

Chapter 8

Summary

Pest management in greenhouses is a challenge primarily due to the conditions that are associated with growing many different types of horticultural crops and the multitude of insect and mite pests that are encountered during production. First, always correctly identify insect and/or mite pests, so the appropriate pest management strategies are implemented. Furthermore, be sure to thoroughly understand the life cycle, biology, and behavior of insect and mite pests in order to target the most susceptible life stages (larvae, nymphs, and adults). In order to alleviate problems with insect and mite pests, a variety of pest management strategies should be implemented, including scouting, cultural control and sanitation, physical control, pesticides, and biological control. Cultural control and sanitation practices are the "first line of defense" in alleviating problems with insect and mite pests, which includes proper irrigation and fertility, and removing weeds from both inside and outside the greenhouse. Scouting provides information on the population dynamics based on the numbers of insect and mite pests present at any given time during the growing season. The use of physical methods to exclude insect pests by installing microscreening will help prevent insect pests from migrating into the greenhouse from outside, which will also alleviate any problems with viruses vectored by certain insect pests such as the western flower thrips. Although there may be a substantial initial financial investment associated with installing microscreening, overall, the long-term benefits will result in a cost savings based on reduced pest problems and decreased pesticide inputs. There are different types of pesticides used in greenhouse production systems including contact, stomach poison, translaminar, and systemic. Some are broad spectrum in regards to activity, whereas others are narrow spectrum. The use of pesticides in greenhouses can be effective in suppressing insect and mite pest populations if they are applied appropriately, which includes timing of applications, coverage of all plant parts, and conducting applications at frequent enough intervals to insure high mortality of the susceptible life stages of a given insect or mite pest. There are a number of factors that can contribute to insufficient performance of a pesticide, including incorrect pest identification, poor water quality, extended shelf life, and using the inappropriate label rate. Furthermore, in order to mitigate resistance, always rotate pesticides with different modes of action. Biological control involves the release of natural enemies, including parasitoids, predators, and/or applying entomopathogenic nematodes. The use of biological control will only be effective if natural enemies are released preventatively, that is, before insect or mite pest populations reach outbreak proportions. Also, always conduct quality control assessments in order to determine the viability of natural enemies purchased from suppliers/distributors. Moreover, understand that the performance of natural enemies may be influenced by environmental conditions (e.g., temperature, light intensity, photoperiod, and relative humidity) and plant types.

Pest management in greenhouses requires a clear understanding of the life cycle, behavior, and biology of given insect and mite pest so that the appropriate management strategy or strategies can be implemented with the assurance for success.

Chapter 9

Suggested Readings

Albajes, R., M. L. Gullino, J. C. van Lenteren, and Y. Elad (eds.). 1999. *Integrated Pest and Disease Management in Greenhouse Crops*. The Netherlands: Klumer Academic Publishers.

Bell, M. L. and J. R. Baker. 2000. Comparison of greenhouse screening materials for excluding whitefly (Homoptera: Aleyrodidae) and thrips (Thysanoptera: Thripidae). *Journal of Economic Entomology*, 93: 800–804.

Berg Stack, L., R. Cloyd, J. Dill, R. McAvoy, L. Pundt, R. Raudales, C. Smith, and T. Smith. 2014. *New England Greenhouse Floriculture Guide: A Management Guide for Insects, Diseases, Weeds, and Growth Regulators 2015–2016*. New England Floriculture, Inc. and the New England State Universities (revised every two years).

Bethke, J. A., R. A. Redak, and T. D. Paine. 1994. Screens deny specific pests entry to greenhouses. *California Agriculture*, 48: 37–40.

Bielza, P. 2008. Insecticide resistance management strategies against western flower thrips, *Frankliniella occidentalis*. *Pest Management Science*, 64: 1131–1138.

Cabrera, A. R., R. A. Cloyd, and E. R. Zaborski. 2003. Effect of monitoring technique in determining the presence of fungus gnat, *Bradysia* spp. (Diptera: Sciaridae), larvae in growing medium. *Journal of Agricultural and Urban Entomology*, 20: 41–47.

Cloyd, R. A. 2006. Compatibility of insecticides with natural enemies to control pests of greenhouses and conservatories. *Journal of Entomological Science*, 41: 189–197.

Cloyd, R. A. 2007. *Plant Protection: Managing Greenhouse Insect and Mite Pests*. Batavia, IL: Ball Publishing.

Cloyd, R. A. 2010. Pesticide mixtures and rotations: Are these viable resistance mitigating strategies? *Pest Technology*, 4: 14–18.

Cloyd, R. A. 2012a. Indirect effects of pesticides on natural enemies. In: Soundararajan, R. P. (ed.). *Pesticides: Advances in Chemical and Botanical Pesticides* (pp. 127–150). Rijeka, Croatia: InTech.

Cloyd, R. A. 2012b. Insect and mite management in greenhouses. In: Nelson, P. V. (ed.). *Greenhouse Operation and Management* (seventh edition, pp. 391–441). Upper Saddle River, NJ: Pearson Prentice Hall.

Cloyd, R. A. 2012c. *Pesticide Mixtures: Understanding Their Use in Horticultural Production Systems. Kansas State University Agricultural Experiment Station and Cooperative Extension Service*. MF-3045. Manhattan, KS: Kansas State University.

Cloyd, R. A. and R. S. Cowles. 2009. *Resistance Management: Resistance, Mode of Action, and Pesticide Rotation. Kansas State University Agricultural Experiment Station and Cooperative Extension Service*. MF-2905. Manhattan, KS: Kansas State University.

Cloyd, R. A. and C. S. Sadof. 1998. Flower quality, flower number, and western flower thrips density on transvaal daisy treated with granular insecticides. *HortTechnology*, 8: 567–570.

Cloyd, R. A. and E. R. Zaborski. 2004. Fungus gnats, *Bradysia* spp. (Diptera: Sciaridae), and other arthropods in commercial bagged soilless growing media and rooted plant plugs. *Journal of Economic Entomology*, 97: 503–510.

Drost, Y. C., Y. T. Qiu, C. J. A. M. Postuma-Doodeman, and J. C. van Lenteren. 2000. Comparison of searching strategies of five parasitoid species of *Bemisia argentifolii* Bellows and Perring (Homoptera: Aleyrodidae). *Journal of Applied Entomology*, 124: 105–112.

Feng, R. and M. B. Isman. Selection for resistance to azadirachtin in the green peach aphid. *Experientia*, 51: 831–833.

Fransen, J. J. 1992. Development of integrated crop protection in glasshouse ornamentals. *Pesticide Science*, 36: 329–333.

Gerson, U. 1992. Biology and control of the broad mite, *Polyphagotarsonemus latus* (Banks) (Acari: Tarsonemidae). *Experimental and Applied Acarology*, 13: 163–178.

Gruenhagen, N. M. and T. M. Perring. 2001. Impact of leaf trichomes on parasitoid behavior and parasitism of silverleaf whiteflies (Homoptera: Aleyrodidae). *Southwestern Entomologist*, 26: 279–290.

Hogendorp, B. K. and R. A. Cloyd. 2006. Insect management in floriculture: How important is sanitation in avoiding insect problems? *HortTechnology*, 16: 633–636.

Huang, N., A. Enkegaard, L. S. Osborne, P. M. J. Ramakers, G. J. Messelink, J. Pijnakker, and G. Murphy. 2011. The banker plant method in biological control. *Critical Reviews in Plant Science*, 30: 259–278.

Jones, V. P., N. C. Toscano, M. W. Johnson, S. C. Welter, and R. R. Youngman. 1986. Pesticide effects on plant physiology: Integration into a pest management program. *Bulletin of the Entomological Society of America*, 32: 103–109.

Li, Z. H., F. Lammes, J. C. van Lenteren, P. W. T. Huisman, A. van Vianen, and O. M. B. DePonti. 1987. The parasite-host relationship between *Encarsia formosa* (Hymenoptera: Aphelinidae) and *Trialeurodes vaporariorum* (Homoptera: Aleyrodidae). XXV. Influence of leaf structure on the searching activity of *Encarsia formosa*. *Journal of Applied Entomology*, 104: 297–304.

Naegele, J. A. and R. N. Jefferson. 1964. Floricultural entomology. *Annual Review of Entomology*, 9: 319–340.

O'Connor-Marer, P. J. 2000. *The Safe and Effective Use of Pesticides* (second edition). Oakland, CA: University of California, Statewide Integrated Pest Management Program, Agricultural and Natural Resources Publ. 3324.

Opit, G. P., G. K. Fitch, D. C. Margolies, J. R. Nechols, and K. A. Williams. 2006. Overhead and drip-tube irrigation affect twospotted spider mites and their biological control by a predatory mite on impatiens. *HortScience*, 41: 691–694.

Parrella, M. R. 1999. Arthropod fauna. In: Stanhill, G., and Zvi Enoch, H. (eds.). *Ecosystems of the World 20 Greenhouse Ecosystems* (pp. 213–250). New York, NY: Elsevier.

Parrella, M. R., L. Stengård Hansen, and J. van Lenteren. 1999. Glasshouse environments. In: Bellow, T. S., and T. W. Fisher (eds.). *Handbook of Biological Control* (pp. 819–839). London, UK: Academic Press.

Pilkington, L. J., G. Messelink, J. C. van Lenteren, and K. Le Mottee. 2010. "Protected biological control"—Biological pest management in the greenhouse industry. *Biological Control*, 52: 216–220.

Sanchez, J. A., D. R. Gillespie, and R. R. McGregor. 2003. The effects of mullein plants (*Verbascum thapsus*) on the population dynamics of *Dicyphus hesperus* (Heteroptera: Miridae) in tomato greenhouses. *Biological Control*, 28: 313–319.

Stanghellini, M. E., S. L. Rasmussen, and D. H. Kim. 1999. Aerial transmission of *Thielaviopsis basicola*, a pathogen of corn-salad, by adult shore flies. *Phytopathology*, 89: 476–479.

Stenersen, J. 2004. *Chemical Pesticides: Mode of Action and Toxicology*. Boca Raton, FL: CRC Press.

Sütterlin, S. and J. C. van Lenteren. 1997. Influence of hairiness of *Gerbera jamesonii* leaves on the searching efficiency of the parasitoid *Encarsia formosa*. *Biological Control*, 9: 157–165.

Tauber, M. J. and R. G. Helgesen. 1978. Implementing biological control systems in commercial greenhouse crops. *Bulletin of the Entomological Society of America*, 24: 424–426.

van Lenteren, J. C. 2000. A greenhouse without pesticides: Fact or fantasy? *Crop Protection*, 19: 375–384.

van Lenteren, J. C. (ed.). 2003. *Quality Control and Production of Biological Control Agents: Theory and Testing Procedures*. Wallingford, UK: CABI Publishing, CABI International.

van Lenteren, J. C., L. Z. Hua, and J. W. Kamerman. 1995. The parasite-host relationship between *Encarsia formosa* (Hymenoptera: Aphelinidae) and *Trialeurodes vaporariorum* (Homoptera: Aleyrodidae). XXVI. Leaf hairs reduce the capacity of *Encarsia* to control greenhouse whitefly on cucumber. *Journal of Applied Entomology*, 199: 553–559.

van Lenteren, J. C. and J. Woets. 1988. Biological and integrated pest control in greenhouses. *Annual Review of Entomology*, 33: 239–269.

Ware, G. W., and D. M. Whitacre. 2004. *The Pesticide Book*. Willoughby, OH: MeisterPro Information Resources.

Wyatt, I. J. 1966. Insecticide resistance in aphids on chrysanthemums. *Proceedings of the British Insecticide and Fungicide Conference*, 3: 52–55.

Xiao, Y., P. Avery, J. Chen, C. McKenzie, and L. Osborne. 2012. Ornamental pepper as banker plants for establishment of *Amblyseius swirskii* (Acari: Phytoseiidae) for biological control of multiple pests in greenhouse vegetable production. *Biological Control*, 63: 279–286.

Yu, S. J. 2008. *The Toxicology and Biochemistry of Insecticides*. Boca Raton, FL: CRC Press, Taylor and Francis Group.

Zilahi-Balogh, G. M. G., J. L. Shipp, C. Cloutier, and J. Brodeur. 2009. Comparison of searching behavior of two aphelinid parasitoids of the greenhouse whitefly, *Trialeurodes vaporariorum* under summer vs. winter conditions in a temperature climate. *Journal of Insect Behavior*, 22: 134–147.

Index

T - #0142 - 111024 - C206 - 234/156/9 - PB - 9780367574772 - Gloss Lamination